STONE UPON STONE

Mary Osborn

To dear Martha
With many thanks
for all your friendship
and support over the
years and with love
from Mary

STONE UPON STONE

The story of the Stanton Guildhouse

Mary Osborn

A late rose blooms upon the thorn
So is my life's endeavour spent
To this one purpose, flowering late
As love's intent becomes at the last
Life's sacrament.

Published in 1995 by
The Guildhouse
Stanton, Nr. Broadway
Worcs WR12 7NE

in association with Green Books Ltd
Dartington, Totnes, Devon

Printed by Biddles Ltd
Guildford, Surrey

ISBN 0 9505225 1 1

Contents

The author wishes to express her gratitude to Mr and Mrs Norman Scoular for their faithful support of the Stanton Guildhouse Trust and their generosity to it over the years. She is also very grateful for the specific bequest from the Scoular family which has made the publication of *Stone Upon Stone* possible.

Foreword

This book tells the story of a remarkable woman and the fulfilment of her dream. The driving force of Mary Osborn's long life has been a deeply held belief that there is a spiritual aspect to even the most simple human activities: especially those that involve craft skills and the use of basic materials. She has always seen crafts as an essential element of the common life, and their revival as a vital component in the regeneration of rural and urban communities. *Laborare est orare* is a text she often quotes, and it condenses her faith as neatly as any three words can. The beautiful Guildhouse at Stanton is a living and thriving instrument of that faith – in craft skills, in community, in the life of the spirit.

Mary Osborn's ideas belong to a distinguished tradition, to which in their different ways Tolstoy and William Morris and Mahatma Gandhi belong. And it was Gandhi, when she met him in London in the Thirties, who more than anyone set her upon her lifelong path. This book is a record of her journey along it: her years at Kingsley Hall and with the Community of the Way in pre-war Bow, her work with the children and village people of Laverton, her development of the concept of a Guildhouse, her struggles to finance and build it, and the work it has done in the twenty-two years since it opened its doors.

I first met Mary twenty years ago. In the late Sixties my wife was a member of the Yorkshire Guild of Spinners and Weavers. Through another member she heard about the Guildhouse in Gloucestershire, then being built, and was shown photographs of its half-completed buildings. A few years later we drove down from Yorkshire for what must have been one of its first open days. We then realised that this was the Mary Osborn who had written an article on Community and the Village in a book we had in our library.

In that article, written over fifty years ago, she described the impoverishment of rural society in those days. She wrote of the need 'for people, whether in groups...or in families or as individuals to settle in villages throughout the countryside, with something of the Franciscan spirit and prepared with all the faith, love and initiative they can muster to assist in rebuilding the broken fabric of the country's life.' Remembering the poverty-stricken villages of my native Suffolk before the war, I understood what she meant. The article also prefigured her dream, the dream she turned into reality: she wrote of her hope for 'something in the nature of a Folk House...where men may learn that work is prayer, service and fellowship. Such a place could provide a centre for the rebirth of craft guilds.'

In the half century since then our villages and their populations have changed, socially and economically, almost out of recognition. Yet the Guildhouse that Mary created still has relevance, and still fulfils a need. There are many hundreds of people whose lives have been touched by the Guildhouse: by the crafts they learned there, by the fellowship they enjoyed there, by the friendships they made there, and by the spiritual nourishment they received from its days of retreat.

In this book Mary Osborn tells in her own words the story of her work, her faith, and the fulfilment of her dream. It will be of great interest to many. It will also be an inspiration to all those who know and love the Stanton Guildhouse, and wish to see it remain as Mary made it.

JOHN SPENCER
28th December 1994

1

Foundations

This is the chronicle of a spiritual pilgrimage, a journey of exploration into the interior self which was also to find expression in the outer material world; the inner inspiration and outward manifestation, stone upon stone, fused together in what was eventually to become a community Guildhouse in the village of Stanton in Gloucestershire. The beginnings of such an exploration lie dimly veiled in the past, perhaps in childhood or even at an earlier state of one's spiritual searchings – who knows? What is certain is that inspired by the relevance of the Christian faith as an answer to the problems of this and every age – personal and collective – a light gleamed and beckoned at the end of a dark cavern. All true inner conquest leads from the darkness of unknowing into the light of God's scrutiny in which everything stands truly revealed.

Something of this light had seeped into my mind in my late teens. It so happened that I came across a book which moved me deeply. It was a character study of Gandhi and Lenin. This book, which assessed the philosophy and influence of these two outstanding figures, made an indelible impression on my youthful seeking mind. For no two leaders could be more diametrically opposed, both in the intrinsic nature of their personalities as well as in their influence upon the course of events in their respective spheres of activity and influence. I was immediately drawn inescapably to the Gandhian way of life with its compassion for the poor, its philosophy of non-violence, its certainty of the power of God which could lead all men into the paths of truth. For had not Gandhi declared that if all men denied God he would be His sole witness?

Even more fascinating to my newly awakened consciousness of the power of Christ's teaching was that Gandhi had been so inspired by the Sermon on the Mount that he had actually woven its tenets into his activities in the field of international politics. He saw in the example of this Carpenter of Nazareth, with His consciousness of oneness with God and His timeless beatitudes of purity of heart, peace-making, humility and long-suffering, the eternal truths which could confound the power and arrogance of the proud and uncaring. The lesson went home and remained indelibly engraved on my soul. To me Gandhi embodied the light of truth in this generation. Christ embodied it for all time. On the other hand, I was mentally shocked by the utter ruthlessness of Lenin's influence as portrayed in this book and on historically documented evidence. Life under the Czarist regime had engendered fearful deprivation among the poor, but the tyranny of a violent revolution was equally appalling, bringing death and destruction in its wake to those who opposed it. For me the impact of Gandhi's philosophy was undoubtedly a milestone in the exploration of my inner consciousness, and I cherished the hope, then a seemingly almost impossible dream, that one day I could meet Gandhi.

Having trained as a Froebel teacher, for there were few opportunities at that time for training in social work, I was determined to share life as closely as possible with people most in need of friendship and support. I had already experienced some of the deprivations which many were called upon to suffer at that time, having worked as a volunteer during school holidays in the welfare centre of a Manchester cotton mill, as well as in a children's orphanage permeated by an overpowering aroma from the too liberal daily use of carbolic soap. One could feel there the chill winds of poverty, despair and homelessness and I wanted to share these experiences and in an idealistic way to bear some of the burden.

To find the place where I was meant to be at this juncture was the next concern; and by a series of events in the summer of 1931 and the sharing of ideas with many friends, links were forged with Muriel and Doris Lester, resulting in an invitation to join them in their social work in Bow, in London's East End. This was a very deprived area with bad housing, high unemployment, with all its attendant problems arising from the inevitable sickness which constantly prevails among the chronically under-nourished. Some years previously, these two sisters had been deeply moved by the spiritual and social needs of this area in which their father owned a number of properties. Feeling the need to give something back to the tenants involved, with the help of family and friends they had set up a centre of fellowship and caring.

It began first of all in one of the ordinary Bow houses and extended to a disused chapel named in memory of their brother Kingsley, later known as Old Kingsley Hall, a building which they rescued from complete decay. Later a new purpose-designed building was erected for the children's and people's work to be known as Children's House. This was inspired, I believe, by an idea evolved in Vienna. The new Kingsley Hall, a third spacious building also purpose-built and providing facilities to meet the many spiritual and social needs of the adult population, had also been recently established when I arrived.

Looking back, the pattern of living in Kingsley Hall can only be assessed in the context of the more open religious orders, where a simple way of life is accepted as the norm and where to give all means to have all, for then the real values of life in work and worship are communally shared. This involvement was eminently suited to anyone in need of a challenge which would stretch him or her almost to the limits of endurance. It was an experience which quickly laid bare one's inner deficiencies. The majority of the twenty or so helpers involved were young and inexperienced. As I have said, there was little organised training available for such work at that time. It was simply a question of being thrown in at the deep end of every difficult circumstance and sinking or swimming for one's very survival.

It was a hard school, but had its own curious and often baffling rewards, for those warm, big-hearted Eastenders were unpredictable, and sharing their joys and sorrows was fraught with uncertainty.

We rose at 6.30 a.m. and were then involved in one or

other of the many domestic chores unless it was our turn to prepare breakfast for the ten or so staff. Then there were morning prayers before breakfast at 8.00 a.m., after which our own rooms were attended to and the real day's work began, correspondence to be answered, the sick or absentees from various groups to be visited, preparation for meetings for which we were individually responsible, children's after-school play hours and evening clubs and classes for teenagers or adults until 10.00 p.m. Sunday was a day of different services for all the various age groups. We had one day off a week and were supposed to have a free hour during the afternoon, but this was often encroached upon by some emergency or urgent need, such as visiting one of our members in the local hospital.

The day usually ended by some of us gathering round the kitchen table for a hot drink and a sharing of the day's experiences, or a lively discussion on some concern of current interest. This pattern was augmented by an occasional visit to the theatre, a concert, a public meeting or a peace demonstration, such as those at the Hendon air display at which we were accompanied by other Christian pacifists. The ploy there was to get inside the enclosure with our anti-war leaflets before the police turned us out. These were always good-tempered encounters; the police were often sympathetic to our cause and almost apologetic for having to do their duty in this way. They were both humorous and courteous.

The children of Communist parents were also there in force, reciting in unison, 'We want scholarships not battleships, we want playing fields not battlefields.' People generally were highly suspicious at that time of anything related to peace or Communism.

We were caught up in other peace demonstrations too, such as the May Day procession which, after passing through the centre of London, made its way to Hyde Park. This was representative of all the main peace groups including the Christian pacifists led by the imposing yet benign figure of Father John Groser, whose church in Watney Street was a sanctuary and an inspiration to all who had the privilege of finding it. When he took charge as priest there, the church was in a very run-down condition, but with the help of a number of unemployed in the neighbourhood, he transformed it into a place of peace and beauty. He himself wove the vestments, altar cloths and frontals. It eventually became a spiritual home to many.

In retrospect life in Bow was a Franciscan experience, with simple food, accommodation and the minimum of pocket money. This was based on the rating received at that time by the lowest-paid member of society – the soldier who, when Kingsley Hall was founded, received only a shilling a day. It was felt by the founders of this charity that we who were concerned with peace-making and reconciliation, and against war, refusing to take part in such conflicts, should also be prepared to accept the same material gain as a soldier. This arrangement was generally accepted, though special and individual circumstances were also taken into account.

The very small remuneration we received also kept us closer to our neighbours whose material plight was sometimes beyond belief. There was no social security, only relief, which had to be sought with deep humiliation, and only when the family was in direst need and there was no other source of help available. This relief usually amounted to sufficient to cover the scandalous rent of half a bug-and-rat-infested house, with a few shillings left over to feed and clothe the family. Little wonder that everything of the slightest value went into pawn – sometimes even wedding rings – so that the family could be kept alive. But in most cases the human spirit soared above the seemingly insoluble problems following in the wake of long years of unemployment with courage and an unbelievable generosity that would put the more richly endowed to shame. Our neighbours would literally share their last crust, and did so when the Hunger Marchers from Jarrow arrived in Bow during the Depression in 1936. Poverty, death and disaster were their constant and in some sense mystical companions which they learned to accept as guests in their midst. The acceptance was probably easier because these were experiences common to all and which no family escaped at one time or another.

My own work covered all aspects of life as time went on, but initially it was an exploration of the areas where I best fitted. I cared deeply about all the many needs, but at that time I was painfully conscious of being young, inexperienced and at times wholly inadequate. The one qualifying factor was the mutual desire for friendship and this applied to us all in a greater or lesser degree. I had previously gained some experience of working with teenage girls through being involved in a Church organisation. In Bow this was a field where some creativity was called for, especially among girls working all day at dull monotonous jobs in factories. Spinning had always fascinated me. So, having arranged for a spinning lesson on a hand-spindle, which incidentally absorbed a whole week's pocket money, I introduced this simple but satisfying craft to some of the teenagers. They were fascinated – using the wool produced for knitting or simple weaving. We also made our first experiments with vegetable dyeing. Eventually, through the generosity of a friend, I became the excited possessor of a spinning wheel, and nothing could have been more welcome as a Christmas gift. This wheel was of course also shared with the group, to their great delight. Activities among the younger children as well as with their parents brought many challenges, as well as the privilege of being counted among their friends. To be accepted was the key to most doors. Years later I visited some of the first groups of spinners, by that time married with families of their own. They recalled those early days with much pleasure.

2

Meeting Gandhi

This was the background against which for me the first miracle happened. I had hardly settled into these new surroundings when in 1931 the news came that Mahatma Gandhi was coming to England for the Round Table Conference, and had chosen to stay among London's poor rather than in the West End. In 1926, Muriel Lester had visited India where she had been introduced to Gandhi by a mutual Indian friend. Through this meeting a link of sympathy and understanding was forged between the poor of India among whom Gandhi was working and the people of Bow. Only the poor can truly share the sufferings of the poor. Once the link was forged the chain remained unbroken, so it is little wonder that Gandhiji, as he was familiarly called, chose Bow as his venue during his stay in England.

The news of his coming broke upon us all so unexpectedly, and I can recall the feeling of rippling anticipation with which it dawned upon my consciousness. So I was to meet Gandhi after all! Was this an answer to the deep desire which over the years had become subconsciously a fervent prayer? What was the underlying purpose of such an encounter? Would he prove to be the strong saintly caring figure communicated to me through the book which had come my way several years earlier? These and many other questions sprang to mind.

Excitement mounted as the time for his arrival drew near. Even the drab little streets vibrated with anticipation. The children, ever ready for some colourful new experience, gathered constantly around our doors with always the same question, 'When is Mr Gandhi coming?' For a time he was to become their personal possession and they would all glow in the light of his reflected glory. Little did they realise the anguish which was bringing him to England – the question of Swaraj, or independence from British rule, in the face of an almost totally unsympathetic government and an often cynical public.

In Bow it was different; the preparations for the great event, though essentially serious, gathered about them a festive air. Accommodation was prepared in the simple cell-like rooms which opened on to the small roof garden of Kingsley Hall. Bedsteads were removed and mattresses laid on the floors in true Indian fashion. Food would be simple, mainly raw and dried fruits, vegetables and goat's milk which the local dairy was able to supply.

The actual day of Mahatma Gandhi's arrival came at last. Before he was to arrive, a special time was set apart for meditation in the early hours of the morning, for we were all deeply conscious of the spiritual dimension of this historic occasion. Then followed numerous interviews of the staff with pressmen and photographers. At 4.30 in the afternoon people had begun to crowd round the doors of Kingsley Hall, awaiting admission to the reception. By six o'clock the large hall used for services as well as meetings and the club-room above it were crowded with members, while representatives from various groups, local clergy and others, waited in the library for a personal introduction to the Mahatma. About 6.30 loud cheers from the crowd outside intimated his arrival. Our Indian friends then made their way up to the first floor library by a back stairway. After a while, the library door opened, and Mahatma Gandhi entered the club-room. Those first impressions of that unique figure one could never forget.

A profound silence surrounded him and out of the depths of that great heart of India radiated the spirit of a saint. A hush fell upon those assembled as the Mahatma passed across the club-room and on to the balcony to greet the cheering crowd in the street below. It was a tremendous moment.

Half an hour later Mahatma Gandhi and his friends joined us for evening worship, which was conducted at our request as it was at the Sabarmati Ashram in India. There was first a period of silent meditation and then, with heads bowed, the chanting of Hindu prayers. This was followed by a further period of silence, when all barriers of caste and creed were broken down in the oneness of our humanity in our search for the realisation of God's presence. It was a deeply moving and unifying experience which we were to share even more poignantly in the days which lay ahead. As time went by, Gandhiji also incorporated into these acts of worship readings from the New Testament, as well as those Christian hymns which meant so much to him. These included 'When I survey the wondrous Cross' and 'Lead kindly Light'.

As far as his own personal prayer life was concerned, for more than fifty years Gandhiji meditated night and morning on the following verses from the Bhagavad Gita, and devoted all his effort to translating them into his daily activities. They are the key to his self-transformation.

He lives in wisdom

Who sees himself in all and all in him,
Whose love for the Lord of Love has consumed
Every selfish desire and sense-craving
Tormenting the heart. Not agitated
By grief, nor hankering after pleasure,
He lives free from lust and fear and anger.
Fettered no more by selfish attachments,
He is not elated by good fortune
Nor depressed by bad. Such is the seer ...

When you keep thinking about sense-objects
Attachment comes. Attachment breeds desire,
The lust of possession which, when thwarted,
Burns to anger. Anger clouds the judgement
And robs you of the power to learn from past
Mistakes. Lost is the discriminative
Faculty, and your life is utter waste.
But when you move amidst the world of sense
From both attachment and aversion freed,
There comes the peace in which all sorrows end,
And you live in the wisdom of Self.

The disunited mind is far from wise;
How can it meditate? How be at peace?
When you know no peace, how can you know joy?
When you let your mind follow the siren
Call of the senses, they carry away
Your better judgement as a cyclone drives
A boat off the chartered course to its doom ...

He is forever free who has broken
Out of the ego-cage of I and MINE
To be united with the Lord of Love.
This is the supreme state. Attain thou this
And pass from death to immortality.

Late that first evening on the day of their arrival Gandhiji and those who had accompanied him from India – his youngest son Devadas, two secretaries Mahadev Desai and Pyralal, and Mira Behn the daughter of an English admiral who became one of Gandhi's most devoted disciples – all joined in the usual Saturday night social time. This weekly event, instituted as a counter-attraction to the local pubs, was immensely popular and affectionately known as Joy Night. And a most joyful occasion it always proved to be – a cheerful meeting-place for friends and for whole families. There were refreshments and usually some special event engineered by members of the staff.

To the delight of all, Gandhiji moved among the assembled company as one of them, talking to old and young while holding the children in his arms. No other gesture was more appealing to these warm sincere Eastenders, and throughout his sojourn they gave him their unstinting friendship and support. He had chosen a most hospitable abiding place for the duration of his stay – materially deprived, but spiritually endowed in abundance. At our Indian friends' invitation we attended prayers at 4.15 the following Sunday morning, and we in turn invited Gandhiji to speak at our evening service. Seated upon the altar steps he spoke to us in slow gentle tones of that power by which he lived and worked. 'Prayer' he said, 'is to the soul what food is to the body, yet prayer is far more important to the soul than food to the body because at times we can go without food and the body feels the better for the fast'. 'We can over-indulge in food,' he said, 'but we can never over-indulge in prayer.'

To drive home his point he continued with this theme for some time, applying its precepts to seeking out the true purpose for our lives, and we all felt the certainty and quiet strength of his own continuing communion with God. His words also shed a light along our own individual paths, and we felt the inspiration of God's presence flooding in upon us through the utter dedication of this devoted apostle of truth, for as he always insisted, 'Truth is God, not God is Truth.' Had he not also declared that if the whole world denied God, he would be his sole witness? The problem for all of us is to be certain of the spiritual thermometer which claims to register truth. The Christian gospel gives us the surest guidelines here in the West, if men could fully understand all its implications in terms of personal, national and international commitment. Gandhi seemed to appreciate the whole spectrum of involvement. For most of us 'we only see through a glass darkly'. For me personally, truth in relation to my own life was to be a long and arduous search, with a great deal of overcoming to be accomplished.

3

Seeking the right path

Perhaps of all the memories of Gandhi's visit, the most memorable were the early morning walks upon which we embarked with him after morning prayers. At 5.30 a.m. a miscellaneous little party set off along the drab streets now enhanced by the starry heavens above and at times a crescent moon. We usually made for the canal, known locally as 'the cut', turning on to the towpath and along by the dark mysterious stretch of water. Some of the inhabitants, hearing that Gandhi took his morning walk at this hour, would stand at their doors to greet him.

In the beginning the party usually consisted of Gandhi and his son Devadas, his two secretaries Mahadev Desai and Pyralal, Mira Behn, and other members of the staff, four policemen and a couple of Scotland Yard detectives, together with any visitors who happened to be staying with us at that time, who were eager to walk and talk with this great little man. Others would sometimes arrive from the West End by taxi, having heard that Gandhi walked at this hour and that there might be an opportunity to talk with him.

Some of these people had come with an honest desire to learn more about India's problems at first hand, others were equally eager to criticise or even blatantly to heckle him, but Gandhi treated them each and all with the utmost courtesy, never raising his voice, nor ever displaying any sign of impatience or irritation. How often were we reminded of the beatitudes – 'Blessed are ye when men shall revile you...' – as we listened and were sometimes also drawn into these encounters. This kind of confrontation turned enemies into friends and was certainly an object lesson to would-be peacemakers.

These early morning walks had been Gandhi's custom for many years, giving anyone who wished to speak with him an easy opportunity for doing so. We also had the advantage of sharing his inner thoughts concerning India's deepest needs. India's problems became our own and we were eager to help. Gandhi himself was very concerned that all misunderstandings should be cleared away, as he claimed that there was colossal ignorance of the whole Indian situation.

On one occasion I enquired what Gandhi's attitude was to Christian missionaries. He said they were really welcome if they came in a spirit of true service, identifying themselves with the people, wearing Indian dress and living very simply with the poorest of the poor. He himself undertook the most menial and arduous of tasks, such as the cleaning of latrines which would normally be undertaken by the untouchables living in the area. He did this as a gesture of identification with the most needy and as an example which he hoped others would follow. Many missionaries, however, because of the life-style they had adopted, were far removed from the lives of the needy. Others realised this and having very successfully bridged the gulf were now accepted as India's own people. We suggested that perhaps a term at his own ashram at Sabarmati might be helpful to those offering to work in India. Gandhi warmly welcomed the idea and said he would be delighted to co-operate in such a way as had already been the case with Dr Stanley Jones and others. After this, revisits could periodically be arranged.

One morning we were joined by an enthusiastic French woman who raised the whole question of education in relation to the caste system. It will be recalled that Gandhi himself had worked out a most comprehensive scheme of basic education, to enable children to play their full part in the life of the village community. This educational programme embraced every aspect and need of rural populations and its influence was widespread. Gandhi said the caste system was the one which deeply disturbed him. It was an evil of Hinduism which must be eradicated by Hindus themselves. In former times caste did not present barriers between people, but each caste was a separate contributing vocation which was established horizontally, so to speak, and worked side by side with others devoid of stigmas. Later the whole arrangement of the caste system was turned over and placed vertically, with the outcastes being the very bottom log in the pile! And so it remained until Gandhi took practical steps to cut across the whole system, renaming the outcastes Harijans or Children of God.

Originally it was deemed wiser for children to follow their father's craft or trade because of a likely aptitude, as well as providing the facilities for gaining first-hand knowledge and experience. It was also thought a sensible arrangement for women to marry within the caste into which they were born because there they would be more at ease. Gandhi thought that the role of every woman was to be a mother, not necessarily physically, but in a spiritual sense. He saw marriage as a noble calling which should be looked upon as a service to country, husband and children. But he was mainly concerned about the dignity of every human being, and he contended that a system which denied or destroyed this essential birthright was working against God's purpose for mankind. For him in later life celibacy became an essential discipline in the fuller realisation of what he called soul-force in the spirit of God finding expression in the world through dedicated human beings. To this end all our faculties and potentialities must be directed.

These thoughts and many more flowed day by day through our discussions. When on one occasion a young debutante, who had chosen to work with us for a time as an antidote to what she deemed a pointless existence, asked Gandhi what he felt to be the real purpose of life, he replied with the unexpected answer, 'To die!' 'To die,' she repeated incredulously, 'but I am young and don't want to die yet!' Gandhi smiled at her response and then proceeded to explain his meaning. He did not of course mean to die physically, and seeing her amazement continued to expound upon the meaning of these words. He contended that it was necessary for us all to die to self in order to live for God and man. One was reminded of Christ's saying, 'Except a corn of wheat fall into the ground and die it abideth alone, but if it die it bringeth forth much fruit ...'

His questioner's reply amused him. 'Well, I think I'll enjoy myself till I am thirty and then I will do what you say.' Interestingly, her life moved in an entirely different direction. But who knows what lasting influence this encounter had made? For no-one could possibly talk with Gandhi without feeling that an ineradicable dimension had been added to life. In the light of this experience, the troubles, conflicts and deprivations of this earthly pilgrimage have a more purposeful counterpart in heaven. There is a cross at the heart of the universe, but there is also a resurrection inherent in every aspect of suffering, and Gandhi's experiences reiterated this truth most poignantly.

A deep and abiding joy had also its place in our sharing life with our Indian friends. On one occasion Gandhi invited all the children in the immediate neighbourhood to talk with him. They were delighted and one can see him now sitting on the floor with a crowd of bright-eyed quick-witted children of all ages around him. They had a wonderful time together, the children plying him with questions and he responding with obvious enjoyment. None of the children present would ever forget that afternoon, but little did they know then how privileged they were to bask in the friendship of one of the greatest leaders of all time!

4

Hand-spinning

On another occasion both Gandhiji and Mira Behn surprised me by showing considerable interest in our recent hand-spinning efforts. Hitherto I had not realised how much a part of Gandhi's philosophy hand-spinning had become, in his attempt to enable the poorest Indians to achieve self-reliance with the help of cotton-spinning among other crafts. The Khadi movement, as it came to be known, developed not only as an exercise towards self-subsistence but also as a meditative experience – shared eventually by people from many walks of life.

Gandhi's interest in hand-spinning was roused into action after his involvement in the revival of handweaving in 1915, but it was not until 1921 that the spinning wheel found a place in the Congress programme. He had discovered its survival after an intensive search, on coming across a widow who was working among the untouchables. It was she who found a few poor Moslem sisters living in Visapur who still practised the craft of hand-spinning. From the moment of this discovery Gandhi saw the possibility of reviving hand-spinning as a vital contribution to his programme of self-help for the desperately needy villagers.

Fired by this newly awakened enthusiasm he did everything in his power to encourage its development as widely as possible, concentrating on the design and production of spinning wheels (charkas) which could be constructed very simply and easily transported. These quickly became widely available together with the small metal hand-spindle (takli) which even young children could use. Gandhi saw spinning as a truly co-operative development to relieve poverty and to promote the dignity of labour. At its peak its revival spread to over fifteen hundred villages bringing with it the all-important philosophy of creative renewal.

It is known that the dyeing of yarn, whether of cotton or

silk, had been practised for centuries, during which time countless intricate patterns had evolved with the use of brightly coloured yarn dyed with plants and other substances of natural origin. Gandhi also contended that the hand-spun yarn need not be dyed because white was composed of all the colours of the rainbow. This was also a kind of mystical conception, expressed in the words of George Matheson:

'Each sees one colour of Thy rainbow light
Each looks upon one tint and calls it heaven.
Thou art the fullness of our partial sight
We are not perfect till we find the seven.
Gather us in.'

To return to our spinning efforts, we were still only using hand-spindles at that time, so Gandhi suggested that we should have the spinning-wheel which had been presented to him by Indian students on his sixty-third birthday. This was an old traditional English wool wheel, given not because it was thought that Gandhi would use it, as he had his own charka or cotton wheel, but as a symbol of the Khadi movement. However, on Gandhi's return to India there was a strong feeling among certain members of the staff that this wheel should be kept as a memento of Gandhi's visit, in the tiny room which he had occupied during his stay. However, during the last war when the bombing of East London was devastating whole areas it was decided that the wheel should be sent to me at Laverton in the Cotswolds, where I was then working and living. By this time it had been badly damaged, but we were able to repair it and to put it to good use again. I wrote to Gandhi, telling him that the wheel had found a home with us at last and was being used by village children. He was delighted, as this was so much more in line with his beliefs than keeping it unused as a museum piece.

So many memories crowd into my mind as I recall Gandhi's sojourn with us. He was always so generous in taking time to respond to one's individual seeking, so when I tried to formulate at his request my own beliefs for his comments and assessment, he replied with the following letter, which I later discovered was his usual custom in response to discussions.

> Dear Sister,
> I have read your letter carefully and heartily endorse all you say. You rightly say that we are one with the universal spirit. I have found that in order to realise this state, we must serve all that lives. This service is only possible if we reduce ourselves to zero. Self-effacement i.e. self-sacrifice is the law of life. And lest we feel it is I who produced a particular result, we must learn to know that no man can ever alone produce a result. We must therefore work without attaching ourselves to results. Ours is to work, the results are in the hands of God. You seem to be on the right path.
> Yours,
> M K Gandhi
> 16.11.31

This letter seemed to sum up the heart of my seeking – the desire to become at one with the 'poorest and lowliest and lost', also to feel myself bound up with the whole of creation, in willing bondage to the universal spirit which is God's love for all mankind.

Gandhi could reach out and experience the heart of different faiths. He, along with practising Christians everywhere, could understand the essence of communion with Christ. This was a source of real inspiration to him in his self-effacing activity in the service of his fellow-men. For Christ manifested God's love in the face of our human condition. He who had trodden this way before knew himself to be one with God, for had He not said, 'I and my Father are one' and to his followers, 'I am the vine, you are the branches'? Gandhi had not said to me, 'You *are* on the right path' but 'You *seem* to be on the right path', and herein lay the challenge – to seek and hopefully to find the right path.

5

After Gandhi

The impact of Gandhi's visit was unforgettable, and most experiences which followed took on a deeper significance and a sharper meaning. Spiritual and moral values were more clearly defined and far more issues presented themselves as being black or white and not just grey. My own inner convictions became stronger and I was prepared to take courage in both hands in fields of spiritual activity where formerly I had been more diffident. Ideas and experiences began to be expressed more readily through poetry.

Encouraged by Muriel Lester I also entered into areas of Christian ministry reserved formerly for priests and others specially ordained to meet such needs, because many of our neighbours were as sheep without a shepherd, owing no allegiance to any church or organisation. They were just outsiders in the eyes of the world, part of the faceless mass of humanity. But to us they were souls for whom Christ died and to whom His love could still be made manifest. He cared what happened to them and so we must care. Besides, they were our friends and our relationships were a two-way traffic, equally valued from both sides.

I remember at this time coming upon a passage in a book by John Ruskin entitled *Unto This Last*. The thoughts the author shared moved me deeply. 'Ye sheep without a shepherd, it is not the pasture that has been shut from you, but the Presence. Meat! Perhaps your right to that may be pleadable, but other rights have to be pleaded first. Claim your crumbs from the table if you will, but claim them as children not as dogs; claim your right to be fed; but claim more loudly your right to be holy, perfect and pure.'

The first experience I had of being led into this kind of shepherding was when one of the mothers asked me to baptise her child. On sharing my astonishment, which was my first reaction, Muriel said, 'How lovely, Mary! If Mrs S thinks you are a fit and proper person to do it, so you must.' And so I did and can recall today the beauty of that child with such wondering eyes who was named Jean Margaret, and signed with the cross of Christ, the first of a number I was privileged to baptise into the fellowship of the Christian faith. There was a very special joy about this kind of sharing, but there were other moments of intense sadness. One of these occasions I can vividly remember still.

It happened when I was asked to take the funeral service for what in those days was termed 'an illegitimate baby' – to country people a 'love child'. He was a delicate baby when born, exceedingly pale and waxen-looking. As was so often the case, he could not survive the vicissitudes of his surroundings, contracted pneumonia and lived for only six weeks. His young mother and the whole family were stricken with grief. The burial was to take place at the vast East London cemetery and never before had I experienced such desolation as when the pathetic little cortège was led to the grave in which there were already the remains of many previous burials, the dead flowers lightly covered with soil. Surrounding the mourners were hundreds of graves, a number of them gaping, almost moaning, like this one, for all the sorrows which had been committed to them, watered by countless tears. There was no dignity there.

And then relief came when the simple, kindly gravedigger appeared on the scene. With bowed head, he lifted the tiny white coffin reverently, gently laying it in the grave. We all found his presence comforting. It was not through the words uttered and the prayers offered, but through the compassion of this humble man that we felt the love of God surrounding us. And so we left this tiny coffin with its sleeping child and cross of flowers in God's compassionate care and that of the gravedigger.

Many will undoubtedly frown upon such blatant unorthodoxy. But who is to say that we are to deny anyone this kind of ministry in their hour of greatest need, especially when they know of no-one else to whom they can turn. 'If ye do it unto the least of these my brethren, ye do it unto me,' said Christ. Perhaps after all we should be less concerned with the outer formalities of our faith which tend so often to imprison Christ in some measure. If Christ's kingdom is to come upon earth, He must be set free to do His work in however small a measure the world over. In this we share St Teresa's thoughts:

'Christ has no body now on earth but yours, no hands but yours, yours are the eyes from which is to look out Christ's compassion on the world, yours are the feet with which He is to go about doing good, and yours are the hands with which He is to bless us now.'

Bereavement was only one of the many spectres which stalked prematurely among our neighbours and friends in Bow. Sometimes long-term illnesses, especially among children, were even more distressing. We tried to counter this by offering first of all a nursery school for the two-to-five-year-olds with a nourishing mid-day meal, medical supervision, good health-forming habits, regular sleep and last but not least, opportunities for learning to live and share together.

Later there were holidays for the children by the sea or in the countryside, and camps for all the upper age-groups including the adults. These were never-to-be-forgotten experiences for all who participated, including the helpers involved who bore the heavy responsibility of running them. I still have recollections of repairing leaking tents by lamplight in the middle of the night, and the recurring daytime hazards which included chasing armies of wasps out of the large stone jam or marmalade jars! Most memorable of all were the times of morning worship – sometimes on the windswept Sussex Downs, the wind rippling through a sea of minute wild flowers, or in a Hertfordshire bluebell wood vibrant with the song of nightingales. This was always a time of honest inner sharing when anxieties were released, and the peace of God settled into our hearts.

6

Attempts at Identification

Meanwhile, long-term unemployment had its devastating effects, not only on the men themselves but upon their whole families. To be unwanted, almost outcasts from the labour market and hence from society itself, left scars which are present with us even today. How often did one hear them say, 'It's work we want, not the dole'? And how many hundreds of miles did they tramp in search of work and all in vain? Historically, memories are long – not only are the sins of the fathers visited upon their children, but their deprivations also.

The real question was, how could one bring a new sense of purpose into a situation fraught with such despair? As it happened an opportunity presented itself. It was a Monday morning, and definitely a blue Monday to the dishevelled group of women gathered in our road. I remember one dear soul who used to wind her thin black hair round a partially toothless comb, finally pushing it into position at the back of her head where it remained nonchalantly throughout the day! She was nodding vigorously when I came upon them, and was consequently drawn into the conversation. The gist of their talk was all the difficulties with which their lives were surrounded – bad housing conditions, the problem of feeding and clothing the children, especially to provide them with shoes, always a major anxiety, almost as high on the list of priorities as food. The pawn-shops, the pubs and the money-lenders as well as the absentee landlords, all came in for a round of impassioned criticism. Finally, where was the sense of struggling at all, they asked? As one of them had once remarked, 'Miss, I am as good a woman as God ever made, but I can't live up to it!'

When this storm of resentment had blown itself out, I ventured to make a suggestion, very tentatively, to the effect that life could be much more purposeful for us all if only we could share our problems with each other a little more realistically and in the spirit of the early Christian Church. If those of us who cared would meet together perhaps once a week to discuss the most urgent problems which needed to be tackled, in the light of Christian teaching, life could be enriched for us all and not least by the act of togetherness itself. There might also be opportunity (and I personally was sure there would be many) for showing our practical concern and caring for each other on a day-to-day need-by-need basis.

Consequently it was through this encounter that a small caring fellowship came into being. Anyone was free to offer suggestions for the subject of our weekly concern. When we met together we made this particular need the theme for our devotions, with relevant hymns, readings and prayers, chosen by the women themselves, who in turn took part in these informal services. Significantly one of their favourite hymns began with the lines. 'I feel the winds of God today, Today my sail I'll lift.'

This happy fellowship not only renewed our lives within but also found expression in all the practical ways that can be shared in a close community. In Bow each street was in one sense a family. The sharing of food had always been an unwritten code and there was also the common uniting experience of birth and death. Now this was extended to the sharing of necessities for the arrival of a new baby or for a period in hospital. If anyone possessed a best anything it was shared gladly.

Soon there followed a most unexpected development. The men, impressed by their wives' new-found enthusiasm and zest for living, asked if they could have a similar gathering. At that time I was inhabiting a minute flat in the oldest of the three buildings. They suggested that perhaps they could meet in my tiny sitting-room for discussion. So the day and time were fixed, although I had no idea what to expect, but one thing was certain – it would be a very different proposition from the women's group. I viewed this venture with a great deal more nervousness, but the fact that the men had asked, that the initiative had come entirely from their side, helped me to muster up a little more courage in preparation for the first encounter.

I had no idea what they wanted to discuss, so it was a question of being led by them. All I could do was to prepare the setting, seeing that the room was warm and comfortable, the chairs arranged – for how many I did not know – also making sure that the kettle was boiling for the inevitable cup of tea. Presently there was a timid knock on the door and on opening it with, I hope, a welcoming smile, a large burly figure stepped inside. Removing his cap, he sat down on the opposite side of the fireplace. He broke the silence with the usual pleasantries. He was a kindly man and fortunately for me he was also extremely voluble, which avoided the embarrassment of too many uneasy silences.

This state of affairs continued for some considerable time, and feeling my concern that no-one else had arrived, he reassured me by saying, 'Don't worry – they're a bit shy. There'll be more next week. You'll see.' Whereupon I made the tea – always a symbol to Eastenders. I sat down again, while my friend with cup in hand began to pour out his life story, feeling no doubt that this was the appropriate procedure. The evening was undoubtedly his, with his story punctuated by me with an occasional nod. There was no breathing space for comment. He left with a similar cheery word of encouragement to the one which had heralded his arrival.

The days passed all too quickly and the appointed evening was again upon us. I went through the same ritual of preparation. This time the knocking was more confident and on opening the door there stood my friend with two more of his pals. We went through the same procedure and sure enough no-one else arrived. Sometimes the conversation piece was reduced to a solo, the speaker replying to his questions with his own spontaneous answers. More often the conversation was a duet, but mostly a trio – all three contestants speaking at once. It seemed obvious that my presence was superfluous apart from setting the scene and making the tea. This augured well for the future; the whole thing would just run itself, for my friends were quick-witted and never at a loss for words.

The next week five or six arrived. Over the ensuing weeks the snowballing process continued until the room was filled almost to overflowing. A number of those present had suffered unemployment over a long period of years, and were continually faced not only with semi-starvation, but with a feeling of utter hopelessness. There was no place in society for them, and worse still, no purposeful future for their children. Nowhere had they found an answer to their questioning.

At first the discussions were for the most part negative criticisms of existing institutions – the Church and its Bishops, the Government and its Members of Parliament and humans in general, with the exception of 'good old George' who was in their eyes a 'true Christian'. This referred to George Lansbury, the much-loved and deeply-respected Member of Parliament in whose constituency we all lived. Some of the men had lately become intensely interested in Communism which was being highly publicised and widely discussed. It was also being espoused by some intellectuals, not for their personal benefit, but on behalf of the poor, as had also happened in Russia. This promotion of dialectical materialism engendered at best an atmosphere of agnosticism and at worst widespread atheism with its attendant denigration of spiritual values.

7

Serious discussions

All these uncertainties had an influence upon the climate in which my Bow friends embarked upon their serious discussions, a course they opted to take after realising that their negative and sometimes anarchistic approach was leading nowhere. Then came the following challenging remark, which sparked off their course in a more positive direction: 'I reckon we'd better do some serious study.' 'What shall we study?' was the next enquiry. 'A big problem,' another commented thoughtfully. At last the light dawned. 'I know,' he ventured. 'How about *Christ and Communism*?' Then turning to me he added, 'That should suit you and most of us.' It seemed an amicable compromise. Fortunately I had read this fascinating book, written by Dr Stanley Jones, an American missionary working in India, who had been deeply inspired by Gandhi. It was only a short while before that Gandhi had stayed with us at Kingsley Hall and had impressed everyone who met him by his obvious sincerity and dedication.

The exciting part about the choice of this particular book was the realisation that for the men it would be what might be termed a rough ride. It would shake their ideas about both communism and Christianity to the foundations. We embarked upon this venture with tremendous enthusiasm. It was inevitably a thought-provoking experience, which very soon took us into deep waters. The main stumbling-block was the men's obvious lack of knowledge concerning Christianity.

At last came a courageous suggestion. 'Well, I reckon we'd better do some serious Bible Study,' came a voice. 'A very good idea,' came another voice, a little closer to earth this time. Was there no-one who would challenge this idea? Yes, of course, dissension was bound to show itself. 'Go on,' boomed a third voice, 'You're kidding! The Bible isn't for the likes of us, it was written by blokes in colleges.' Never,' countered several voices in unison. There was a surprised ripple of laughter and then we got down again to the serious business in hand.

'What do *you* suggest?' asked the first spokesman solicitously, directing this question towards me. 'I think it's a very good idea,' I replied, and suggested we began with the gospel of St Mark, which is the earliest record, based, it is thought, on a first-hand account of what happened, as recollected by the disciple Peter.

Half-trusting, half-sceptical, they approached the next session with the Bibles which I had been able to muster in their hands. The initial atmosphere of tenseness and uncertainty soon gave way to a more relaxed approach. Familiar passages and parables, forgotten since childhood, were obviously re-read with pleasure. The somewhat sensitive passages such as 'Use a little wine for thy stomach's sake' (1. Tim 5.23) were eagerly ferreted out and flung at me with the mischievousness of schoolboys. In spite of the good-natured chaffing, the deeper impact of this 'serious study' was making inroads into their consciousness, to such an extent that when we reached the end of St Mark's Gospel they were eager for more.

The most logical corollary seemed to be the Acts of the Apostles, and after some discussion it was decided that we should explore the influence and effects of Christianity on the lives of ordinary men and women, as recorded in the Acts. This study was another exciting experience, summed up in the following culminating sentiment – 'If only people would live like that today, there wouldn't be any problems.'

'Why don't we try?' I ventured. 'All that believed were together' – such was the foundation upon which we were called upon to build. Herein lay an answer – a life of love and brotherhood in community. Some of us would work towards that. Bow had so many environmental limitations that we felt we must move out to the country where we could sow, reap and rejoice in honourable toil, where we could create things of beauty and usefulness with our own hands, and where together we could share a life of communion with God. This life would not in any sense be an escape from reality. But we hoped it would offer in embryo an answer for us to the gravest problem with which the world is faced today, the art of learning to live together peacefully and creatively.

8

Dewdrop Inn

It happened that a number of the men had never been outside their local area, so plans for an occasional outing together evoked considerable excitement. One such occasion was a visit to Hampstead Heath which proved to be the first of several, for we made friends with a shepherd and his wife who tended a flock of sheep on the Heath. Another visit was closer to home but even more exciting. This was to the Dewdrop Inn in Vallance Road, Whitechapel, formerly an old pub but converted into a unique social centre by Mary Hughes, the indomitable daughter of Judge Hughes, the author of *Tom Brown's Schooldays*.

The contrasting styles of life were remarkable, for the Hughes' family home in Mayfair had also been the venue for eminent friends interested in socialist ideals. These visitors had included Lewis Carroll, Charles Kingsley, Robert Browning and many others. Their concern for the poor had so impressed Mary Hughes that she determined to give her life to their service, which meant forsaking her wealthy heritage in order to live in their midst, and as simply as possible.

Mary was a Quaker and had very strong views on alcohol, smoking, vivisection and war. She was also an ardent vegetarian. It was therefore not surprising that visitors to the Dewdrop Inn were confronted by a window full of posters relating to all these causes. Mary always gave us a warm welcome but when we were snugly settled in her crowded kitchen she would proceed to express in no uncertain terms her views on one social problem or another. One of her favourite stories was of a former heavy smoker who on giving it up was able to buy a small cottage in the country with the money saved, and there he and his family lived happily ever after. Her dissertations always had a moral which she hoped would be taken to heart and acted upon by her hearers.

Meanwhile Mary lived as closely as possible to the people around her, sharing their trials and tribulations as well as the facilities, including the bath, which the Centre had to offer. She cared for everyone who called on her for help. Her food was also very simple – bread, margarine, onions, cabbage, tea and cocoa without milk. This was her main diet; to those who were privileged to be her friends her indomitable spirit was always infectious.

A slightly more ambitious adventure was a day's excursion to Derbyshire, and preparations for such a momentous occasion obviously called for a good deal of generosity and tact in the face of so many material needs. Long-term unemployment impoverished every aspect of life. It was important to all the men to hide their deprivations and appear tidily dressed in the eyes of the world. So there was a great deal of consultation and many exchanges of best jumble sale clothing, so that all the families' presentable garments were equally distributed. Finally the excited party set off on a very early excursion train from Euston to Stockport where we arrived about mid-morning.

So far the journey had been uneventful, although interesting to those who had never previously travelled so far. They did not know that on arrival at Stockport a surprise would be awaiting them. Members of my old church in Wilmslow (where I had lived before going to Bow) had undertaken to meet our little party with several cars and take us for a run into Derbyshire. This experience was a deep breath of fresh air to anyone living in London, for at certain times of the year it was enveloped in a shroud of yellow smog.

We drank in the peace and beauty of the hills and wide open spaces. Then on to Wilmslow where my family provided a welcome meal before the return journey at midnight to London. We arrived there in the early hours of the morning tired but relaxed and happy.

9

Towards Community

One of the most significant developments preceding the tragic outbreak of World War II was the upsurge of renewed interest in community living. Over the centuries this ideal was motivated by the way of life adopted by the early Christians when 'all that believed were together', sharing life at every level of experience – a life-style inspiring to many, but difficult to achieve and maintain. In the past this spirit had found expression mainly in religious communities, where a common faith and discipline provided a framework within which a shared experience could function and ensure some measure of continuity into the future.

There were numerous exceptions, of course, where religious persecution broke the sequence of events. Nevertheless, it was monasticism which kept the light of faith and culture alive during the so-called Dark Ages in Europe, and which has survived to this day, adding a depth of spiritual experience as well as a wealth of theological learning and insight. In 1936 the community movement in Britain was greatly inspired by the arrival of a German community known as the Bruderhof or Society of Brothers. This had been founded by Eberhard Arnold together with his family and friends in an attempt to establish a Christian pacifist rural community. When it became impossible to maintain such a life in Germany because of the rise of Nazism, they moved first to Liechtenstein and finally to England. Here they re-established themselves in Ashton Keynes in a beautiful stone-built farmhouse with a number of spacious outbuildings, readily convertible into extra living accommodation for the many English families and single people who joined them.

Having settled in, the community set up a printing press and a wood-turning workshop to augment their income from other sources. Meanwhile this deeply sharing Christian experience attracted considerable interest. There was also an air of romanticism about their simple life-style, their traditional peasant clothing and the sense of being part of an extended family – even if one was only visiting to help them for a short time. One indelible memory is of the whole community standing in a circle out of doors on a moonlit night, singing old German songs, especially the unforgettable German pilgrim song with its beautiful haunting tune.

Experience of another kind of religious community came about through my friendship with some of the remarkable Benedictine nuns at Stanbrook Abbey and I shall never cease to be thankful for their friendship and understanding, as well as the benefits derived from the pages of numerous books illuminated by their shared thoughts and spiritual insights. I felt privileged when Sister Hildlith, the printer and director of the Stanbrook Abbey Press, renowned for the quality of its craftsmanship and the content of its publications, agreed to print my slender volume of poems 'Flowers on the Thorn', a title inspired by a gift to me by the nuns of a 'Christ Thorn' (euphorbia splendens); this is a viciously thorny shrub from which Christ's Crown of Thorns is traditionally reputed to have been made, and upon which tiny red flowers like drops of blood continuously appear.

Another religious community with which I had some personal contact was the Anglican Community of St Francis at Cerne Abbas. This grew out of the concern of its founder Brother Douglas from as early as 1921, for the unemployed, homeless men who tramped the countryside in search of work and for some meaning for their lives. Brother Douglas called them wayfarers. By 1931, having interested others in his deep concern for the underprivileged, he and they had established themselves as the nucleus of a religious community, taking vows of poverty, chastity and obedience. This community centred round the inspired leadership of Brother Douglas, and had a significant influence on many other people anxious to live more simply in order to share their own resources with those deprived even of basic needs. Consequently, a third order was established, consisting of married and single people living and working in the world at large who wished to support the work which the brothers had undertaken. Brother Douglas became a good friend of our small community in Bow, often arriving at our door looking like the Pied Piper, followed by a crowd of children attracted by his brown habit and kindly bearing.

Apart from the religious orders, of which there were many others, enclosed or working in the world, there have been numerous attempts at community living on purely secular principles, where men have become conscious of the blatant exploitation of their fellow men and felt called upon to rectify this state of affairs even in a limited way. Both of the above patterns of community – the religious and the secular – found expression in those which were newly formed immediately before the outbreak of war in 1939. Most of them were land-based, primarily to absorb the conscientious objectors who were ordered by tribunals to undertake work on the land as an alternative to military service.

Inspired by Leslie Stubbings, whose commitment to voluntary work in this newly awakened movement was remarkable, a community service committee was convened, of which I was a member. Its early meetings were held in our small sitting-room in Bow, where at that time we were also making our own tentative experiments in community living. This liaison committee published a regular broadsheet with news of all the groups involved in what was then becoming a widespread movement. It was also responsible for two significant publications – *Community in Britain* and *Community in a Changing World* – which gave a vivid overall assessment of the patterns of thought and living which were emerging at that time.

However, few of the land-based communities survived for very long after the war ended. But in recent years there has been a renewed interest among young people in moving out to remote country areas, to attempt a life of self-subsistence – a few families living and working together to achieve this idea. As one young friend expressed it: 'This is not a rebellion but a rejection of the hollow materialistic values which many of the older generation have sadly engendered.' What the future holds for these young people is impossible to predict, but while the human heart has a propensity for striving after perfection, this quest will continue, fuelled also by Christian tenets and by the other great religions of the world.

So the summer months of 1937 saw us searching for a home in the country to house a little company of twelve – five children and seven adults, including Judith Honegger, recently arrived from Switzerland to work in Bow, and myself. Inspired by the early Christians who were known as the people of the Way, we decided to call our new fellowship The Community of the Way. The idealism behind this venture was strong enough, but the practical problems involved proved insurmountable at that stage, and after several abortive attempts to find a rural setting into which to move, we were forced to re-examine our immediate needs and spiritual direction. All the uncertainties involved during this period of searching proved something of a dark night of the soul, but nevertheless we did receive a measure of 'ghostly comfort' and were undoubtedly sustained by Lady Julian's message:

He said not
Thou shalt not be tempted
Thou shalt not be travailed
Thou shalt not be afflicted
But He said
Thou shalt not be overcome.

Eventually came the realisation that the weakness lay not in the ideal but in ourselves – perhaps we were not yet ready to embark upon a venture which demanded so much. With this realisation our despair gave way to a sense of deep thankfulness. We felt we were being directed to work out the ideal more fully in Bow before attempting to establish a rural community. The vital question was how and where, and seeking a solution to this immediate problem became our most ardent prayer.

This was answered when Doris Lester generously offered us the use of one large room and two smaller ones in a house she owned in the neighbourhood. Here a new chapter began to take shape and involved more families anxious to take part. Their support was a constant source of encouragement and found expression in countless practical ways, from plastering walls and redecorating rooms to supplying us with a host of spring bulbs to be planted in tubs and window boxes – a sure message of hope.

We knew that our community programme must include time for prayer, study and work, together with the development of a number of crafts to supply our needs. Eventually we hoped to provide a livelihood. They would also offer each individual opportunities for creating from within themselves some special and unique contribution to the life of the community as a whole – so important to the spiritual health and well-being of those whom society has made to feel faceless and of no account.

As the days passed our large room became a hive of activity with spinning, weaving, dyeing (on an open brazier), carpentry, wooden toy-making and later printing and shoe-repairing, all taking place in various corners, to say nothing of the inevitable preparation of meals! It was a truly Franciscan experience, and apart from sporadic problems and the worry of breaking into our last pound until a further commission was completed, most of our days were permeated by joy.

The quality of our craftsmanship was eventually enhanced by the arrival of Herbert, a German refugee who one day just arrived on our doorstep, having been directed to us by some Quaker friends. Herbert was a surgeon by profession who, to his cost, had objected to the persecution of the Jews. He was also an excellent woodworker and printer. Under his guidance and instruction we bought a second-hand press and forthwith set up a printing corner – undertaking small commissions for paper-headings and later producing our own Christmas and Easter cards with original designs and verses.

Meanwhile discussions with the children were always a thought-provoking experience – their choice of topics ranging from 'Did men come from monkeys?' to 'The existence of God'. On such an occasion one of the boys remarked, 'I know God is there – sometimes I can feel His breath.' Another observed, concerning primitive man's search for God, 'Well, they saw the corn growing and felt themselves growing and they knew there must be someone who was making them grow.'

As time passed, it became obvious that Herbert would never be able to use his medical skills in this country without qualifying all over again. He was also anxious to make his fullest contribution to the needs of suffering people in a wider context. We made various enquiries on his behalf and consequently he offered his services to the Methodist Missionary Society for work in China, where he and a Chinese doctor were put in charge of a country hospital near Hankow. Since then he has had experience of other medical work in Ethiopia and has found fulfilment in many varied spheres of service. Happily we are still in touch.

Our life was considerably enriched by the arrival of Comrade, as he was known by everyone. In his own words he was the black sheep of a self-respecting, hard-working family. He had really lost his way, owing partly to some disconcerting but nevertheless endearing traits. For the past twenty years his only home had been the uncharitable shelter of a 'doss-house'. He possessed hardly a stitch apart from the clothes he stood up in, until we were able to purchase a new suit for him in a sale and collect together from friends other necessary garments; likewise to find him dentures and glasses, of which he was also in need.

Encouraged by Herbert to use the facilities offered by the local Council, Comrade found his way at least once a week to the public baths, returning with a face positively shining from the purifying action of corporation carbolic soap. And so he became a new man. He told us the most incredible stories of his journeyings and of his friends (many of whom were bishops, politicians, rebels of the first water, including suffragettes with whom he was convinced he was tied to the railings in Westminster!). In spite of his inability to extricate truth from fable, Comrade had a really good heart and did everything in his power to help us. He even made a valiant attempt at producing a crop of rhubarb in an old tub. All in all he was a veritable Brother Juniper! 'Now I am going out to hear God's birds whistle' were his parting words as he sallied forth to chop firewood.

Eventually it was possible to find two rooms in the houses of a kindly neighbour where Bob, another unemployed man who was alone in the world and who had recently become our expert toy-maker, and Comrade could make a home together. On the great day of the removal Bob brought all his worldly possessions on a small handcart. But Comrade, living in a state of perpetual Franciscan bliss, brought all of his in a paper carrier!

Our craft-work, apart from stimulating the creative and imaginative powers of our group, made a substantial contribution towards helping our frail bark to keep afloat financially. In the beginning we tried to set a fair price to our productions, but this was not very satisfactory and certainly did not seem to be in accord with the principles underlying the rest of our life. We therefore tried to evolve a charitable principle and to discover a more satisfactory way of distributing our goods. Finally it became quite clear that we must produce for use and not for profit, and in doing this must be prepared never to let money stand between what we could produce and people's real needs. We even extended this idea to a system of bartering our goods for vegetables produced by unemployed men in Kent. This method of distribution demands both discipline and discretion, but at the same time holds many pleasant surprises in store.

10

Shared Experiences

A good deal of serious interest was shown by our many friends and well-wishers in what was for us a newly emerging way of life. Many saw this as a more realistic attempt at sharing than trying to house and integrate a miscellaneous group of people under one roof. They were kindly critical of our long-term intention to move to the country. They spoke of the need of the big cities and of the work we could do if we remained in London. We appreciated this point of view, for one cannot live in the East End of London for years without understanding this side of the picture. Nevertheless, long experience had also revealed the futility of trying to patch up the old system. The task before us was that of working out together an alternative way of living and also, we hoped, the discovery of a more abundant life.

When Christmas came, we prepared to share the festival together. On Christmas Eve we clad the ugly corners of our workroom in festive raiment – green fir and holly branches (sent from the country) and a host of crimson candles. The Christmas tree too was adorned with tiny candles and sheltered beneath the lowest branches was a simple crib, the symbol of God's love for man. A lantern shed its light on the crib. Christmas Day dawned. The meal shared, the carols and the Christmas story round the crib, the lighting of the candles on the tree for friends in many corners of the world, the children's faces, the gifts beneath the tree and the love which brought them thither, and the peace which crept into our hearts at the day's end – these would remain forever in our memories.

Amid joys and difficulties, order and chaos, plenty and poverty, a new life was outwardly struggling to birth. What now remained was to plant this newly emerging cell into a rural environment, where close contact with creative living could nourish it and help it to grow to maturity, even as the religious orders grew. But this was not to be. We had striven together throughout the preceding years, through every means at our disposal from the whole upsurging Community movement, through Church services, meetings, demonstrations, articles and also through new approaches to day-by-day living and sharing, to stave off the threatening ogre of war – but our efforts had been in vain, or so it seemed then. The one lesson which mankind is slow to learn is that God cannot stop war if men desire it. When the majority of the world's population has a deep and total commitment in their hearts to peace-making, only then will war be eradicated from human society. Sadly, that halcyon day seems far distant.

When war finally broke upon the world, our sad and bewildered fellowship was scattered – the women and children being evacuated to safer surroundings and a number of the men, previously unemployed for many years, suddenly found themselves to be useful to a government at war. There was nothing left for Judith and myself to do but to find another context in which to share our common life. Along with countless others we also cherished the hope that, unlike the 1914-18 War, this one would be a conflict of considerably shorter duration and with far fewer casualties. How erroneous our hope proved to be!

After making provision for the remainder of the group who were neither evacuated nor re-employed at that time, Judith and I set off with packs on our backs during the bitter winter of 1940. We hoped to find a corner of the countryside where we might be of use to the inhabitants in the spare time left over from earning a living by spinning and weaving. We also hoped eventually to pick up the threads of our community experience in Bow.

11

War Clouds Gather

I have not travelled widely or often over the years, but the journeys I have been privileged to make have provided significant bridges into the wider world, providing insights into the pattern of several different cultures, their social structures, and also the impact of their political structures.

The first of these experiences was an invitation to Fanø off the coast of Denmark to take part in the IFOR (International Fellowship of Reconciliation) conference in 1939. This was in many ways an exceedingly sad occasion, for the threat of war was overshadowing the world. I found it very hard to find words of encouragement when asked with other young people to lead the morning and evening devotions on one of the days. One thing became evident as the days passed. We would remain faithful to our ideas and to each other in spirit whatever course our various governments would take. The Germans were deeply troubled about the Nazis, and delegates from Estonia, Latvia and Lithuania were equally apprehensive about the possible domination of Russia over their territories – fears which sadly proved to be justified.

Our final act of worship and commitment to each other came on the last day when we all took part in a candle-lighting ceremony, each lighting our own small candle from a larger one, at the same time pledging ourselves to the cause of peace-making whatever the consequences. It was a deeply moving act of spiritual committal, and at the close many of those present, and especially the Germans, were in tears.

After the conference a friend and I decided to hitch-hike to Copenhagen, where we arranged in faith to meet other friends at 9 a.m. the following morning, a feat about which there was some scepticism. However, we set off in high spirits and were picked up by the driver of an empty cattle truck. He installed us in the place of cattle and took us on our way for many miles. When he arrived at his destination he set us down and we were shortly afterwards picked up by a Danish MP and his wife who took us to Ringsted, where they generously invited us to supper before taking us to the Youth Hostel where were were able to stay for the night.

We left Ringsted at 7.15 the following morning and with the help of more lifts arrived at the hotel to meet our friends at 9am, much to their surprise. We were then taken to visit the main sights of Copenhagen, and during the days which followed we were generously provided with overnight hospitality by Danish friends we had met at the conference. This enabled us to see a good deal of the countryside, its villages and towns with their fascinating architecture, and most interesting of all we were invited to visit some of the Folk High Schools where the courses were designed to inspire the students to take back the essential aspects of Danish traditional culture into the life of their villages.

Another excursion took us by train from Copenhagen to Elsinore where we visited the castle and Hamlet's grave, an old age pensioner's home and the famous People's International College whose founder, Peter Manniche, had contributed so much over the years to international understanding and goodwill. There were many other memorable experiences – one recalls with pleasure the sight of the King riding on horseback through the streets of Copenhagen and being greeted as a friend by many passers-by, including of course the children. This is a symbol of democracy from which other countries could learn, while the clear light of the landscape seems to reflect the clarity of vision of so many Danes, who really seem to understand what life is all about.

12

Journey's End

After returning to England our journeying lasted for six months, the time being broken here and there by some unexpected adventure or to help friends who were working with refugees from Austria. During this period we covered a great many counties and, with the help of friendly cars and lorries, an incredibly long distance in miles. Our travels at this time were unaided by signposts, which had been removed as a safety precaution in the event of an invasion. This national predicament led to many baffling situations when we literally did not know where we were. It was even dangerous at that time to be seen studying a map by the wayside. We could easily be mistaken for spies, as one thoughtful policeman reminded us!

As the days passed, we became alive to the fact that never again might we have such an opportunity for studying at first hand the village life of England, the conditions under which country folk live and work, their urgent daily needs, and what is most interesting of all, their philosophical approach to life. It also confirmed our belief, which had already been borne out during the preceding years of work among the unemployed, that the revival of crafts had a great deal to offer in helping mankind to rediscover the creative motivation at the heart of life; and could be a vital means of bringing about reconciliation between man and his creator, between man and his environment. Surely one of the seeds of the present-day Green Movement.

The ideal of creating a rural centre with our Bow friends was never far from our minds in all our searching, but soon the truth was borne upon us that with every available corner taken up with evacuees from the towns, both space and rents were at a premium. So for the time being we were forced to abandon the ambitious idea of finding accommodation which could adequately house a community comprised of several families together with a number of single members. For a self-subsisting undertaking it would also be necessary to have sufficient land suitable for cultivation and also the facilities of a large communal workshop, as well as the necessary funds to launch the scheme. This would be a matter of faith. For Judith and myself personally, wandering from village to village, even with the kind of purpose we were pursuing, gave us at times a feeling of rootlessness. So when our search came, as it seemed to us, miraculously to an end, we were deeply thankful.

The cottage in the Cotswolds into which we finally moved a little apprehensively – for we had no previous connections with Laverton – was the smallest and probably the oldest in the village, with little more accommodation than one and a half up and one and a half down. The upstairs boasted a bedroom and a landing, which became another bedroom and later also housed a large cupboard, bookshelves and several spinning wheels. My bedroom likewise accommodated spinning wheels, bookshelves and a large Danish loom – whereupon the village children remarked, 'Miss, wherever can you sleep?'

Downstairs was a beamed sitting-room, with an old-fashioned open range, with hobs, the fire contained at almost waist-level. Adjoining there was also a large larder, to be stretched eventually beyond its normal capacity, as storage space for a printing press, woodworking tools and a cupboard containing vegetable dyes, as well as a chest of drawers for cutlery, table linen and sundry other necessities – all, let me add, neatly packed away.

The downstairs 'half' was a lean-to outside, with sink, cooker, washing boiler and behind it a tiny bathroom with the essential conveniences except that there was only a cold water supply which had to be heated in the washing boiler. Outside was an ancient shed, which eventually became the village 'cattery' when our sacks of wool for spinning were installed. There was no need for a cat flap as the antiquity of the shed provided plenty of entrances and exits. At the end of the garden path was the relic of an earth closet which finally collapsed with age.

It was some time before the cottage was sufficiently in order for us to contemplate the arrival of our possessions from store. Meanwhile we made a temporary home with a friendly, welcoming village couple with whom we felt instantly at ease. They initiated us into the ways of the village, giving us a vivid biographical account of each family, also their respective orders of merit in relation to the village hierarchy. They introduced us to an uncle who was the village builder and who became the source of all our decorating and repairing needs. We set to work with a will, for under this tiny span of thatch a new chapter was soon to take shape.

Any lurking doubts or fears were quickly dispersed by the children who gathered round the door, day by day, to watch and encourage our labours, for with the help of paint and white distemper the transformation of our newly found home was taking shape. Finally, to the children's never-ending delight it became a veritable Hansel and Gretel cottage. In all our ploys the children volunteered a helping hand and soon became our loyal friends. The arrival of our few possessions was the next item of interest. Seeing the van outside, the children came again to survey the scene. One of the younger boys did not approve of our natural unstained furniture and came some weeks later to tell us so. 'Isn't it time you painted your furniture?' seemed to him a timely comment. But the spinning wheels became the real centre of attraction. They met with whole-hearted approval and great curiosity.

13

Laborare est Orare

Then came a request from one of the girls, 'Please will you teach us how to spin?' Of course we were willing, for it was just this kind of response for which we had been hoping and even preparing. Very soon a little group of the older girls were learning how to spin and this was the beginning of the village spinners and weavers guild. They took their new-found occupation very seriously, gathering quantities of wool from the hedges, decorating their distaffs and spindles with symbolic designs, and learning all the processes (including dyeing with plants gathered locally) by which the raw material becomes a soft, durable and attractive thread. In an incredibly short time the girls were producing quite good yarn which they made into useful articles and garments.

Then one day came a request from one of the boys, 'Please – when are we going to begin?' A suggestion that they might design and make simple wooden toys was greeted with enthusiasm for, being wartime, all toys had disappeared from the shops. To become toymakers inevitably meant to have a place where we could learn to become craftsmen, which led to an untiring search for a suitable workshop. We soon found the place for which we were looking – an upper room in the house of a friendly farmer, approached by an outside stone stairway, in the neighbouring village of Stanton. It was here that the toymakers guild came into being.

We worked together, encouraged by friends who contributed tools, equipment and books for our library. We held guild meetings, adopting the Benedictine ideal *Laborare est orare* (To work is to pray) as our inspiration, and shared with the children our dream of a village Guildhouse to be used by anyone who needed it in the days to come. In the meantime, Judith and I kept the whole thing solvent by making handspun and woven garments and furnishings for interested friends who were anxious to have them; our small but productive cottage garden also helped to sustain us.

A year later we held our first exhibition of handspun knitted and woven garments and gaily painted toys, which included farm equipment and numerous farm animals painted to represent the various breeds of pigs, cattle and sheep as faithfully as possible. The exhibition stimulated considerable interest, especially among our neighbours of all ages, some of whom were beginning to understand the underlying feeling of community expressed outwardly in terms of craftsmanship. These two dimensions – the inner spiritual inspiration with its sharing of life through an unseen but vital communication, and the outward expression in terms of acquiring a high standard of craftsmanship – had been inherent in the foundation of the medieval Guilds, whose spiritual home was the Church.

Our first exhibition was followed by another memorable sharing shortly afterwards when all the children gathered at the cottage. It was the night before Christmas Eve. This time the firelit room was decorated with green boughs and red candles and there were tiny candles shining on the table and set among the toys which the children had made.

We remembered together the very first Christmas through carols and stories. Each child lit a candle and in silence thought for a moment of someone he or she especially wished to remember. When they had all done this, two candles remained, and one of the children said, 'Let's light a candle for peace,' and another said, 'Let's light a candle for all the people in the world.' With the spectre of war shedding its horror upon us all, the silence which followed our candle lighting was profoundly moving. Children can so often penetrate the heart of a situation and this insight was to be manifested again and again during the years which were to follow. When the silence was broken the toys were taken from the table and the children went out into the starry night, each with a precious treasure to be given to a younger brother or sister on Christmas morning.

Very soon, with the ever-increasing activity, our work-room at Stanton began to burst at the seams and it became necessary to look for more adequate accommodation. This was not easy, for most of the available space was naturally occupied by the evacuees from the cities. However, after a thorough search, we found the ideal place – a large timbered room, with a separate entrance, in the house of one of the oldest inhabitants, Thomas Nash, who became our guide, philosopher and friend for four years.

14

Granpy

Thomas Nash, or 'Granpy', as his numerous grandchildren affectionately called him, was a true son of the soil, knowing many of its secrets and bearing its peace in his heart. He had lived with poverty all his days but he never became embittered by the struggle for existence for himself and his family. In his day it was twelve shillings a week and a family of ten to feed and clothe. Nor was the farmer for whom he worked beyond getting him out of bed on a moonlit night to do the haymaking. The days were often so long he had to dig his own garden by lantern light. Granpy's wife also worked in the fields when she could, and took in washing which she did in the evening, yet she still found time to make the children's clothing by hand.

In many ways we were sad to leave our previous Guildroom which had so generously been put at our disposal, but with the help of the children and a handcart we soon settled happily under Granpy's hospitable roof. He took a great interest in all our activities, advising the boys on their designs for farm implements and the accompanying 'stock' – pigs, sheep and cattle. His approval was summed up in the comment, 'It's a good thing these youngsters be learning to work together. That's as it should be.' Granpy was kindness itself in every possible way, providing extra storage space for our materials and encouraging all our endeavours. He also liked to give us little surprises such as baking the largest apples he could find for our lunch.

In spite of wartime rationing we were able to heat the Guildroom by a communal effort, whereby each member would bring a block of wood or a lump of coal for our open fire. The same kind of sharing also made it possible to provide a meal for everyone when we arranged some special gathering. This willing co-operation engendered a warm feeling of community. I often wondered about the long-term effect of the war upon the children and tried to counteract some of its most disintegrating aspects by a more positive approach to our everyday living. Glimpses into the way their minds worked were often very revealing. On one occasion two of the children were discussing walking down to the village shop in the dark, with all its attendant fears, when the nine-year-old girl rounded off the conversation with 'I keep telling him, Miss, there's nothing to be afraid of, because there's no-one worse than yourself.'

It was owing to Granpy's generosity and to the easy availability of our new Guildroom, as well as to the crying need of people everywhere for the reinstatement of positive Christian values, that the work grew considerably. An excellent craftsman, Arthur Henshaw, volunteered his services and with his help a small woodworkers guild for the men and the older boys came into being.

At this time we also acquired a beautiful antique flax wheel, probably from central Europe, and we made our first experiment with the printing press, producing our own greeting cards. Another craft, introduced by Elsbeth Weyel, a German artist friend, was puppet-making – culminating in the construction of our own theatre and the production of plays, the most popular being the story of St Nicholas. This was based on scenes from the life of the saint, with a grand finale when he appears as the traditional Santa Claus complete with sledge and a sack from which he produced small gifts for all the children in the audience, much to their surprise and delight.

By now the various groups occupied the greater part of our time, but there were spaces in between for friends who wished to stay with us and breathe again the peace and freedom of the beautiful Cotswold countryside. There was also time to share our neighbours' needs and the worried concerns of our evacuee friends, adults and children.

We had our own problems too in presenting our Christian pacifist point of view in a world at war. Some of our neighbours were baffled by our stand; but having friends in many countries did not mean that they suddenly became enemies simply because our respective governments were at war. Others were more understanding and supported us whole-heartedly, while a few confessed that they were of the same opinion, but dare not say so openly for fear of losing their jobs, which in many cases meant their home also. Times of war are exceedingly difficult for soldier and pacifist alike, for each contends that he is following his conscience. The complete abolition of war is the only ultimate solution but, judging from present-day events, that day is still far off, although one must never cease to believe it possible.

During the late summer of 1945 Granpy sadly had a fall from which he never really recovered and he left us very peacefully on September 12th. We all suffered a great sense of loss with his going, for we counted him amongst our dearest friends who shared our joys and sorrows, successes and failure, as if they were his own. The children brought bunches of autumn flowers which we twined into a wreath for his resting-place. The simple service in the village church was a fitting farewell to one who had been so faithful in life that death had no fears for him.

Granpy's going meant also the loss of our beautiful Guildroom, as the house was to be occupied by new tenants who needed the extra space. We hunted in vain for another vacant corner and it was at this juncture that the dream of a permanent Guildhouse began to impress itself upon our consciousness. It had been with us for many years as a dream but was now urging to be born as a reality. We took the first essential step in that direction – opening ourselves up fully to God's leading.

Meanwhile we were faced with the problem of moving from Granpy's house all our equipment and materials for which our good friends at Stanton Post Office kindly offered storage space. We then paused to assess this new and unexpected turn of events. For the present we just had to be prepared to wait quietly, prayerfully, and with a willingness to follow immediately when the appointed time came. Meanwhile the work itself must go on.

Some of the groups returned to the cramped accommodation of the cottage, others were provided for in the homes of our members. But these arrangements involved for us all endless complications of constantly moving equipment as well as furniture, challenging all our resources of time, energy and good humour. It was very much like living and working on a constantly moving roundabout, but we were soon to discover that even a roundabout has a still point at its centre and that is where things really being to happen.

Suddenly there was a breakthrough. It was heralded by the exciting news that the whole near-by village of Stanton, at present in the ownership of one family, was to be sold – the houses individually, some with vacant possession; others were to be offered to the sitting tenants at a reasonable figure. There would also be a few vacant plots of land for development, subject to permission from the Planning Authority. The whole village vibrated with excitement, everyone wondering how everyone else would be affected.

Then the brochures with details of the various properties began to appear and information of what we had previously only seen from the outside became available to all. We viewed several houses but none seemed suitable for our purposes, even if we could eventually raise the necessary funds, so we applied for a site and, after literally years of waiting and after all the existing properties had been settled, we were offered almost the last available plot. It was only years later that we fully appreciated the treasure of a site with which our patience had been rewarded.

15

The Wounds of War

At the end of June 1950, Judith and I had the privilege of attending the International Fellowship of Reconciliation Conference at Woudschoten, Holland. The themes of the conference – 'The Christian in an Atomic age', 'The East-West Question' and 'Christian Non-Violence and Peace-making' – helped to clarify one's own attitude to some of the many complex problems which confronted the Christian Church. Nearly two hundred members (representing twenty nationalities and many denominations) were present. The aspect of the conference which meant more to us personally than anything else was the opportunity for the renewal of old friendships and the forging of new ones. These experiences seemed to take on a deeper meaning as the years passed.

Although it was already five years after the cessation of hostilities, the experience of the war itself had left still suppurating sores upon the souls of many Dutch people. Their sufferings had been almost intolerable and many of them only survived by eating tulip bulbs. Many too were also hiding and feeding Jewish friends throughout the war at the peril of their own lives. My friends Kees and Beatrice Boeke who founded the famous school in Bilthoven had many tragic stories to relate.

In Amsterdam an old man told us in halting English of the breaking of the blockade when allied planes dropped food from the skies which seemed like veritable manna from heaven. In Arnhem we heard too from eye-witnesses of the tragedy of the 4,000 British paratroopers who, out of 9,000, lost their lives through a military blunder; one wondered, recalling such incidents, at the cheapness of human life in warfare and the blasphemy which calls upon God to uphold it. Only the spirit of understanding which leads to peace-making and reconciliation can heal the nations' wounds.

We left Arnhem for Cologne late one evening and, owing to a mistake on the part of the travel agents in London, we arrived at our destination at 2.00 a.m. It was too late to make arrangements for accommodation, so we decided to spend the rest of the night on the station. We sat on our cases in the entrance hall and soon discovered that many other people were doing likewise – and not just occasionally, but nightly. These were obviously the homeless of Cologne who could find no other shelter and whose homes had been destroyed by the incessant bombing. Some were old people who carried a little wooden stool and setting themselves down against a pillar, tried to sleep; others were young and crippled, having lost one or even two limbs, and all night long the station swarmed with children. A friendly policeman was constantly on patrol; he smiled to see two Englishmen, likewise stranded, lying prone upon the floor on newspapers, their Trilby hats covering their faces. About 4.00 a.m. one of the restaurants on the station opened, and we were able to sit more comfortably and enjoy a hot drink. At 8.00 a.m. we left Cologne for Heidelberg.

The depression which had inevitably settled upon us on experiencing so much distress, deepened considerably as the train sped through one devastated town after another. The destruction that one witnessed was beyond anything that our imagination had anticipated. It was quite unbelievable that man in his most vindictive mood could be so ruthless. The train was packed so we were obliged with many others to sit on our suitcases in the corridor for practically the whole journey. This, like every other train in Germany, had two coaches reserved for war-injured people, young and old, and it was in the corner of one of these coaches that we found ourselves. We arrived in Heidelberg in the late afternoon and there the feeling of despondency lifted a little. There was no destruction to be seen and the beauty of the old town, nestling among the woods above the gracious river Neckar, gave back to us something of the real Germany. Heidelberg was the centre of the American occupation, and one saw much that was distressing. Nevertheless Prof. Karl Balser and his wife, with whom we were invited to stay, helped us to find the good things which still remained.

They were unforgettable days and each evening we heard the most beautiful music in the Chapel of the old castle or in the Church of the Holy Spirit, and I experienced again that otherness in life which breaks in upon human disasters. The truth and beauty of God and His peace find a vehicle for their message even in calamity.

From Heidelberg we travelled along the beautiful Neckar valley to Biberach, where we had planned to visit an ex-prisoner of war (and his family) who, along with others, had been in a camp at Stanton and had visited us at the cottage in Laverton when they were allowed to fraternise with English families. After their return to Germany life was very difficult and we kept in touch by sending parcels. The reunion was a joyful occasion and the journey itself from Heidelberg to Biberach will long be remembered for the beauty of the countryside, the peaceful villages with their mediaeval architecture, the well-cultivated fields and the orchards laden with fruit. This seemed a paradise indeed after the devastation of Cologne.

At the conference in Holland we had also met for the first time Pastor Rudi Daur and his wife, who very kindly invited us to spend a few days with them in Stuttgart before leaving Germany for Switzerland. This again was an unforgettable experience. Pastor Daur's church, which at that time ministered to 11,000 souls, made tremendous demands upon the spiritual resources of its pastors, and we were very impressed by the way in which their self-giving was reflected again in the life of the Church. It was a rare privilege to share communion with members of the congregation and to partake of a second symbolic act whereby everyone joined hands and sang together a verse embodying the inner meaning of communion. For, as Pastor Daur said afterwards, there is no true communion with Christ without fellowship one with another.

We attended the morning service which followed and were delighted to find the church full to overflowing. It was a most inspiring service led by Pastor Daur and Dr Maas of

Heidelberg who preached the sermon. Dr Maas was the first German to be invited to Israel after the war, in recognition of the courageous work he undertook on behalf of the Jewish people in Germany. It transpired that 32,000 Jews were hidden and fed out of their meagre rations for ten years by German families, in the face of Nazi terrorism, 12,000 of these in Berlin itself. We kept coming across instances of this kind of courage and faithfulness to the highest values of life.

While in Stuttgart, we also visited St Leonard's Church which had been almost wholly destroyed during the war, but instead of demolishing the whole building, the portion of the walls which had survived had been skilfully incorporated into the reconstruction of the present church. Adding to the effect, there hung a large wooden crucifix upon which rested the most peaceful figure of Christ I had ever seen, an unforgettable symbol of hope and resurrection.

During our travels we saw of course only a very small corner of Germany and one which was recovering fairly well, but in the industrial areas the situation must have been vastly different and far more grave. Probably the most significant aspect of the inner resurrection which was taking place lay in the fact that a number of people inside the Church who suffered considerably for their pacifist principles were becoming after the war the spiritual leaders of the people. Being strengthened and inspired by their sufferings, they accepted them not in a spirit of martyrdom, but as a spiritual discipline of the most testing kind. This also applied to pacifist leaders in many other countries. In speaking with them at Woudschoten one appreciated the real significance of the words to be found among the apocryphal sayings of Jesus, 'If thou hadst known how to suffer, thou wouldest have been able not to suffer.'

From Germany we travelled on for a short stay in Switzerland; a number of impressions stand out in bold contrast to the devastation experienced when travelling through the war-scarred cities of Germany. Switzerland was completely surrounded by countries at war and suffered a certain measure of deprivation. But preparing itself for this eventuality, it had deposited vast quantities of food in storehouses cut into the mountain sides, so there was little evidence of serious shortages. In fact the impression was of undisturbed prosperity. On travelling around, the majestic silence of those towering peaks was breath-taking. The first sight of the Eiger, Munchen, Jungfrau and the Matterhorn with the rosy evening glow lighting up its snowclad summit was a never-to-be forgotten experience, with quiet valleys below and mountain villages above with their flower-strewn alps where only the cowbells or an occasional alpenhorn broke the silence... Heaven itself could hardly be more beautiful.

Memories also crowd into mind of the little mountain village of Kippel where Judith and I stayed during the feast of Corpus Christi, with the burgeoning altars of wild flowers set against the mountain scenery at every turn of the road and our own finding of wild lilies on the eve of the festival. Then the day of the feast itself, the procession of the villagers in their national dress, the solemn music and crowded services in the tiny church and before the flower-decked altars, and perhaps most memorable of all the sense of peace which crept into our hearts.

I recall too a visit to the first Pestalozzi village at Trogan and then the days spent in another remote valley with a kind motherly soul who, being also inspired by the spirit of Pestalozzi, shared her home with children of many nationalities from the war-stricken countries of Europe. For me this was to be the last visit abroad for many years to come. As the Guildhouse dream came more clearly into view it demanded most of my time and energy.

16

Pastures New

It was not until 1953 that the whole transaction for possession of two of the most beautiful acres of the Cotswold countryside was completed and put into Trust, with the intention of erecting upon it a village Guildhouse. This was made possible by the unbounded support and practical generosity of an old friend, Florence Rowbotham, who had always encouraged my aspirations in the Bow days and was still at hand to make possible this latest venture.

Although somewhat alarmed at her first encounter with the site, her response was characteristically magnificent. The first impression was of a long-neglected orchard, overgrown with thistles, nettles and bramble bushes – these interspersed with ancient fallen fruit trees. The trees which still remained standing were suffering from years of neglect, so the prospect was not initially promising. However, higher on the hillside the scene opened up extensively, commanding the most magnificent view of hills and valleys and through a gap to the faraway river Severn with the Black Mountains beyond. My friend's summing-up of the situation was, 'Well, if you have the courage to take it on, I will support you.' She forthwith bought it, and continuing to travel with us 'the second mile' she also sent a further thousand pounds to start off the appeal.

A local committee which had been functioning for several years in preparation for this event was now urged on to greater activity since the project was now considerably closer to realisation. Trustees were appointed to hold the land in safe-keeping for the future benefit of the whole community. Then came the strenuous uphill struggle that extended over several years in the effort to make the necessary preparations for the beginning of the actual building work. One of the major deliberations had been the preparation of our terms of reference, to be embodied in the Trust Deed. These involved the intentions for which the Guildhouse was to be established, and included the provision of facilities for the practising of the main traditional rural crafts, together with the extension of those activities which could benefit the spiritual, physical and mental well-being of people of all ages. It also laid down its ecumenical and non-discriminatory basis, implying an open door to the whole world and its needs. The Trust Deed, which had to be in line with the Charity Commissioners' requirements, was finally settled by Counsel and the Stanton Guildhouse Trust became a registered Charity. This was the first hurdle overcome and it provided us with certain tax and rating reliefs and related benefits, but the start of the building operations was still a little way ahead.

There are times in life when two lives essential to each other's development touch for a while and are again separated by outward circumstances but remain inwardly linked. Living in two countries often raises personal and material problems, as so many of my friends from abroad have experienced. Then comes a point at which some decide to return to their native land. Others never return – with or without regret – while the rest strike a balanced compromise, coming and going between two worlds. So it happened at this juncture that Judith, with whom I had worked for many years, and who had contributed so much to the early chapters of the work, returned to Switzerland to take up social work in connection with the Church. It was a sad parting of the ways, and yet, probably, an essential experience for us both.

At this juncture dawned also the inevitable realisation that I would have to face the building of the Guildhouse alone – and yet not alone. My faith was secure. My belief would manifest itself in the days which lay ahead.

To return to the needs of the outer edifice, we had to find the means by which, stone upon stone, the whole structure could be brought to completion. In our efforts to find a way, we were encouraged to appoint a professional appeal organiser. This we did for a short period, but although eminently successful in other charitable fields, he found it next to impossible, even with the support of an impressive list of patrons, to raise funds on our behalf. The main argument against such a venture was that with the welfare state now firmly established and with the accompanying social security, there was then no need for voluntary efforts of this kind, because everyone was provided with adequate services. Succeeding events have however proved it otherwise.

Our failure to solicit public support, except in a very small measure, led to lengthy deliberation about our next line of approach. Finally my co-Trustees decided that I would personally have to take on the responsibility of raising the necessary funds – of all tasks undoubtedly the most onerous as far as I was concerned. My experience in Bow had inculcated a detachment, even a disdain for money – an attitude from which I was to receive a rude awakening with the realisation that without money, the Guildhouse would remain forever a purely idealistic dream. Now there was no option left but to shoulder this burden with energy and imagination. An appeal brochure was prepared and circulated widely. What next could be done to establish our ideas more firmly in the minds of the man and woman in the street? The answer came. 'Go to the Metropolis, of course.' This was not just an invitation but an undeniable command.

So having made initial enquiries with the help of our appeal organiser and gained all the necessary permissions – to spin on Tower Hill (by courtesy of the Keeper of the Tower), on the steps of St. Paul's (by courtesy of the Dean and Chapter), in front of the Royal Exchange (at the pleasure of that august Authority) and having gained overall sanction from the Metropolitan Police, I took my spinning wheel to London and embarked on this unusual pilgrimage. The year was 1961.

17

A Spinning Venture

It all began one Monday morning when, having got my wheel, basket of wool, chair and leaflets safely to the speakers' pitch on Tower Hill, I sat down at random and began to spin. Not a very salubrious spot; it was early and the area had not yet been swept. The usual 'characters' were already occupying their favourite positions along the wall, while others stood around in small groups, hotly debating some controversial topic. The rest patronised the near-by refreshment stalls.

Visitors to the Tower, including numerous parties of schoolchildren, stopped to watch the spinning-wheel and to ask questions. The more ambitious among them accepted my invitation to 'have a go' on a hand spindle. The apparent magic of drawing a single continuous thread out of a bundle of unwashed fleece never ceased to fascinate the onlookers. 'Is all wool spun like this?' asked an oldish man with the eagerness of a child. Technicalities called for constant explanation, and many men confessed that they could never understand how the spinning wheel worked. They were now glad to know.

As morning wore on, the small groups of men who habitually met there to rend each other's arguments and ideologies to shreds were also drawn towards the spinning-wheel. They merged and mingled with the growing midday crowd. We exchanged glances. Then words began to flow, slowly at first. 'And what are you getting out of this?' enquired a man in a cocked hat. 'Nothing in the sense you mean,' I said. 'I have been doing this work for many years, first with unemployed men and their families, and now in the neighbourhood where I am living – trying to build up a creative fellowship.'

'Who cares about fellowship?' retorted my critic. 'Everyone's out for self these days. As for that,' he continued, pointing at the spinning wheel, 'It's escapism, that's what it is. You won't get nowhere with that. It's Government what's wrong.' Thereupon he harangued the crowd with fervour, exhorting them to come to their senses and put the right people in power. The crowds continued to increase until those few square yards of everyman's land were soon packed to capacity. By this time arguments were flowing freely and discussions concerning my own surprising activity reached a more serious level. Many understood the real implications of this gesture, others didn't. 'But you can't compete with the machine,' said one onlooker. 'We are not trying to,' I explained. 'This is not a commercial venture. Just an attempt to add something constructive to the basis of good living. The spinning-wheel is a symbol of all that man can create from his own inner spiritual resources.'

At this juncture, a tall intelligent-looking man drew nearer to press home to the crowd the point about creative living. 'I couldn't agree with you more,' he said. 'Humanity is at the crossroads. Either it will destroy itself completely, or it must rediscover its true life. But you have a hard task before you.' This sentiment was voiced again and again by people from many different walks of life.

After the midday crowd had departed, a young Communist from Stepney joined me for further discussion. He was worried about the continual conditioning of human beings from the cradle to the grave. Presumably, to his way of thinking, only non-Communist countries are guilty of this error! 'People must be free to think,' he said. 'But no one thinks, they just follow blindly like a flock of sheep.' In his mind, what needed changing most was the environment. I suggested that this was not all. We needed a rebirth of spiritual values. He did not get my meaning, and was dubious of anything he termed 'mystic'. The passing crowds paused to watch the spinning-wheel and to join our discussion. During the afternoon I found myself defending before this ardent follower of Marx not only the re-establishment of traditional crafts but Christianity at its source, Schweitzer, Gandhi and the whole philosophy of non-violence. I pleaded also for friendship, tolerance, and sweet reasonableness between individuals and nations, for the very purpose of our life is to create. Destruction is the negation of our whole existence.

The inner struggle upon which we were both engaged finally broke through to a clearer understanding, and turning to my former critic the younger man said, 'You leave her alone. The lady's all right. She's doing a practical job, and what's more, she's fond of children. She's all right.' I felt his final approval had been hard-earned, but I was nevertheless moved by his generosity.

The crowds that were drawn to the spinning wheel on the steps of St Paul's the following day did not challenge my purpose so fiercely. They came not only from many parts of England but seemingly from every corner of the earth to see this crowning glory of Wren's achievement. 'Why are you here?' they asked. A question for which I was amply prepared.

There was genuine approval of the whole idea. A village Guildhouse to meet the needs of rural dwellers for creative fellowship. And residential too, so that others can share the inspiration of our 'family' from time to time. 'Family' is how one of the village children expressed it. Many of us exchanged addresses, looking forward to our next meeting, perhaps at the Guildhouse.

It was, however, outside the Royal Exchange that the discussion reached its deepest level. In the heart of the city one felt an uneasy depression, an obvious sense of frustration and disillusionment, an anxious concern about the present chaotic patchwork of living, and the still more uncertain patternless future.

'There seems to be only one answer,' I said. 'Man must build up a tremendous force for good by patiently trying to rediscover those things which belong to his peace. This demands a positive way of life, a forward-looking intent, for we can never stand still. If we are not creating we are destroying.'

'But we are all caught up in the same senseless net and there seems no way out,' said a businessman. 'If only we could return to a simpler way of living, how much happier we would be. If we could return to this,' he said, and his hand came down emphatically on the spinning-wheel. At this moment the whole venture in London had turned full circle, and when I visited Tower Hill for the last time my former critic in the cocked hat also greeted me with undoubted friendliness. Then turning to the crowd he said, 'Do you know, this lady's got a good idea, a very good idea. She wants to build a community for sanity.' So he had got the point after all.

Many other efforts were to follow. These included renting a stall in the market-place at Stratford-on-Avon, using it as an exhibition as well as a talking-point – the whole effort being supported by some of the older girls who had been learning to spin at the cottage in Laverton and whose finished garments were also on display. They too understood the true purpose of our endeavour and could converse freely and knowledgeably with interested visitors to our stall.

The Town Hall Square in Leicester also proved to be another interesting venue for our appeal effort, and this was easily arranged when I received an invitation to speak to several groups in the area including a Church Fellowship and a meeting of the local Inner Wheel. When she heard of my intention, a friend also kindly offered to accompany me. As I began to spin, a crowd of mainly unemployed teenagers gathered round to watch and to discuss the purpose of this unusual event. Some were sceptical but others quickly understood what this demonstration was all about, and between the two groups a lot of good-natured bantering ensued. Then suddenly the unexpected happened: one of the crowd who had been carrying a canvas under his arm revealed that it was a completed picture which he was now offering to sell in aid of the proposed project. This announcement created a mood of subdued astonishment and then the fun began as they tried to determine its value in terms of money. Finally a price was agreed upon and it was purchased by an enthusiastic art student to whom, as I heard afterwards, it continued to give much pleasure. This was just one of the many surprises we encountered as our venture slowly unfolded.

From time to time the kindly head gardener and his wife at Hidcote Manor would turn out their potting shed so that we would have space to arrange exhibitions there, especially on a public holiday. Such events created considerable interest and a little financial support, encouraging us to spread our wings far and wide in the hope of stimulating interest in other places. However, it was in Cheltenham that our efforts on behalf of the proposed Guildhouse met with the greatest response, with the exception perhaps of the London venture which was a special experience in its own right.

18

The Vision

The venue in Cheltenham was the largest store, Cavendish House, whose manager we had prevailed upon to allow us the use of a bay in one of the departments for a whole week. Having agreed he did everything in his power to make the event a real success. We staged an interesting exhibition of craftwork contributed by members of our various guilds, with spinning demonstrations to add life and movement to the scene. We also distributed literature and were able to discuss our aims and ideals in depth with many of the casual customers to the store – emphasising the need for such centres of creative fellowship, especially in the rural areas.

Visitors to our exhibition responded with interest and concern, so I ventured to appeal through the local press for building materials as well as money. Miraculously, a response came almost immediately from a local builder, Philip Marshall, who offered us five old Cotswold stone barns which were about eight miles away from our site. He said that we could have the stone as a gift if we would demolish the barns and completely clear the site in readiness for a building development scheme.

When I first saw these barns and stood there in the midst of them, I knew a miracle had happened, and that the words of an old prophecy were, for me, coming true:

'The vision has its own appointed hour,
It ripens, it will flower;
If it be long, then wait,
For it is sure, and it will not be late.'
Habbakuk 11: v.3 (Dr Moffat's translation)

Incidentally, during the preceding years, we had acquired thousands of beautiful old Cotswold slates as they became available, much to the amusement of some of our friends to whom we appeared to be providing a roof for a building without walls! But we knew that slates would become increasingly difficult to obtain in any large quantity when we were eventually in need of them, and this proved to be so. However, the builder's generous gift had put the walls also 'on the horizon' so to speak, as well as yielding a further quantity of sound stone slates. It now remained to find helpers able and willing to demolish the barns and transport the stone to the site.

This kind of provision, like so much else, appeared miraculously when two men, Dave Twinberrow and Don Taylor, seemed to fall more or less from heaven at the moment when they were most needed. We employed them at their requested modest wage; they demolished the barns and transported, load by load, four hundred and fifty tons of stone to Stanton by lorry. Had we not accepted the offer, this stone would probably have disappeared into road-making, so we looked forward eagerly to giving the one-time beautiful old barns a fitting resurrection.

Meanwhile, plans for the actual building were being prepared on a professional basis by Iorwerth Williams, an architect friend who had lived in Stanton for some time. His wife had also been in charge of the village school which was a flourishing addition to life at that time, putting a youthful heart into an ageing community. All our possible future needs had to be taken into account, as well as the initial vision, while our precious plot itself was being cared for most solicitously by numerous volunteers.

It will be recalled that this was a badly neglected orchard on a hillside overlooking the village, itself a poem in stone. During the years of waiting before the foundations could be laid, the village children had so often accompanied me during their free time from school, to cut thistles and nettles and generally help to tame the wilder aspects of the undergrowth. We had picnics too, which were fun, and always an enticement to work on 'the land' as they called it. From the architect's point of view, the main objection to the site was the lack of a plateau on which to build, especially when I insisted the building should be situated in an elevated position on the hillside. He contended that a more reasonable siting would be close by the track at the foot of the hill, because of the problems involved in negotiating the hillside with materials which would also require the construction of a further extended track. 'But how sad,' I responded, 'to be denied the healing peace and beauty of the wide expansive view which could never fail to inspire and lift one's spirit.'

However, this dialogue was finally settled and the inherent problems quickly remedied when Roger Hanks, the owner of a local quarry, came to assess the quantity of stone which would be needed to make negotiable the existing badly rutted farm track which was our only means of access. 'Where are you going to build?' he enquired. 'About two thirds of the way up the slope,' I replied. 'But there is no flat area on which to build,' he objected mildly. 'No, not yet,' was my response, 'but surely we could just move the hillside by pushing the earth across to the far side of the shoulder.' 'Yes, let's!' he responded, with the kind of conviction that makes one feel that the job is already done. 'We have a large bulldozer up at the quarry and we'll bring it down.'

And so the deed was accomplished. Within the next few days, the large mechanical animal arrived on the scene and with its huge jaws moved the mountain. It also levelled out a further track to bring the appointed site within reach of the activity which would constantly come up from below as the building grew. When the architect next visited the site, I could not determine whether he was dismayed or relieved, but now the work could really go ahead in earnest. So it was, after all, to be set in a 'high clean place' and with God's help it might one day also become a small 'city set on a hill, that could not be hid.' The world had need of such sanctuaries.

I recall a paragraph in our appeal literature at that time which reiterated this belief. 'There is no brittle idealism in the conception of the scheme. It does not seek to recreate the past. It is a vital forward-looking intent to create what the late Eric Gill, craftsman and writer, called "a cell of good living in the chaos of our world" with everything that such creative effort implies.' The finalising of the plans, however, required still further consideration so that the building would sit comfortably on the hillside, while enhancing the surrounding countryside and at the same time providing adequate facilities for the creative needs of all who would eventually share its life and work.

19

Work Begins

At last the plans were completed and permission granted by the Local Planning Authority. Early in 1963 the foundations were pegged out in readiness for the first spadeful of soil to be lifted. Now all that was required was the labour. For some months previously our application for a team of volunteers had been considered by International Voluntary Service, a non-political organisation inspired by its founder, Pierre Ceresole, whom I had met in Bow many years previously. I had always been very impressed by this movement, which was conceived as an army of young people – largely students – dedicated to peace-making, international understanding and helpfulness wherever needed. They asked for no payment apart from food and lodging and were prepared to tackle the most laborious and often exceedingly heavy work.

The founding of International Voluntary Service for Peace in 1920 was met with an overwhelming response which has been steadily maintained over the years. There must be countless people in many lands whose lives had been shattered by disasters of all kinds, who now have reason to be profoundly thankful for the practical help and compassion they received from these young enthusiastic volunteers.

We eventually heard that our application had been considered sympathetically and that a team of ten volunteers would be arriving in August. In April, with the help of Dave and Don, the work began in earnest, with further excavations into the hillside to provide for the three-storied wing of the building, thus allowing for more working space under the same roof span. It would also give the building the appearance of naturally growing out of the hillside.

The evening before the work on the foundations was due to begin, I came to the site alone and the words of Louis

Untermeyer came to mind:

I would rather work in stubborn rock
All the years of my life
And make one strong thing
And set it in a high clean place
To recall the granite strength of my desire.

I prayed for strength and courage to see the task through to completion.

While the work was going on through the wettest spell of the year, a spacious hut, delivered in sections during the winter and buried in the snow for weeks, was finally retrieved and erected in readiness for the arrival of the volunteers. Inside the hut provision was made, along with all the necessary equipment, for cooking, eating and recreation. Washing facilities were erected behind the hut; a field boiler was acquired and several tents for sleeping. The Red Cross had generously arranged to loan us blankets. The palliasses we made ourselves and filled them with hay supplied by a friendly farmer.

In all these preparations we were helped by two other volunteers who offered their services simultaneously – a young German girl from Berlin and a very young elderly Quaker of seventy who amazed our village friends by sawing up a large quantity of wood to fuel the field boiler. They were also duly impressed by her exceedingly sparse vegetarian diet, a discipline she had laid upon herself, along with the Spartan habit of taking a daily cold bath!

It was still raining when the first team of ten IVS volunteers arrived at the beginning of August, but after the first fortnight, which was cheerfully spent digging trenches and laying drainpipes and constructing manholes, conditions improved considerably. Then a really good spell of weather enabled the volunteers with the help of three local men – one of whom gave his services that summer – to push the work into action with a leap. They did an excellent job and by the time the team left the greater part of the drainage system had been installed, a trench dug for the water supply, the footings in the basement had been reinforced efficiently, the first layer of concrete had been laid over the basement floor and the footings dug out round the whole of the building. I was then commissioned to lay with due ceremony the first block in the basement as a foundation stone.

After the volunteers had left, the local men continued with the building, bringing two of the basement walls up to ceiling height, as well as completing the damp-proof course throughout the whole building before the bad weather set in. Then the work had to be brought to a halt until the following spring.

Meanwhile, the local groups continued to meet mainly at the cottage, now with a renewed spirit of anticipation, having put so many initial problems behind us. As for the children, their faith was boundless. 'But you won't forget that we were the first volunteers, will you Miss?' said one of them when the IVS team had arrived. How could I ever forget?

While the work at the cottage was continuing, our slender resources were running out and it was imperative to make a great effort to solicit further support towards the building of the Guildhouse. So I sent copies of our appeal literature, along with particulars of some press publicity we had already received, to more than two hundred members of a well-known organisation whose interests and ideals were closely allied to our own. Disappointingly at that time, only one person responded with any kind of conviction. However, he proved to be the one person we needed at this juncture, for he in turn introduced us to the Director of a charitable foundation with considerable assets, and prevailed upon him to visit us. Incidentally, I had approached the same foundation previously but without success.

After his visit, the Director persuaded his Trustees to stretch their umbrella a little and take us underneath, which they sensitively did to the tune of two thousand pounds. This substantial support was tremendously encouraging, not just in material terms, but in the realisation that there were, after all, some people whose ideas and aspirations were in tune with our own, and who did not feel that the welfare state provided for every need and eventuality. This insight was reflected encouragingly in a letter we received from the Director after his visit. He wrote, 'You clearly have many difficulties still in the way, but faith conquers mountains, and you are establishing yourself on a mountain with that wonderful view across the valley, and I feel sure that you will succeed in your endeavours and bring renewed courage and hope to many in an ever-widening circle.'

In a struggle such as this, when one attempts to translate an idealistic dream – or maybe a vision – into the more substantial fabric of stones and mortar and everyday living, such words fall upon the spirit like manna from heaven. This encouragement brought with it a sense of exhilaration and a quiet inner certainty that all would be well.

In the long spell of wonderful summer days in 1964, building operations re-commenced and continued without let or hindrance. For five months volunteers from many different countries lived and worked on the site, cheerfully learning the hazards and intricacies of building construction under the most able guidance of two skilled men. The stone-mason, Jim Powell, helped occasionally by his brother Jack, began the very skilled and demanding task of constructing the outer stone walls against the inside skin. The traditional way of Cotswold building is to fill the cavity between the two walls with the chippings from the stones as they are trimmed for laying in the various courses. We used a technique that is termed 'random rubble building' which enables one to use varying sizes and shapes of stones. Here and there a much larger stone, termed a jumper, is inserted. This enables the mason to change the depth of the remaining course if necessary. This is an ingenious method which facilitates the utilisation of a great variety of shapes and sizes of stones, eliminating a great deal of wastage. The variety also adds considerable beauty to a well-constructed wall.

The summer of 1964 saw considerable progress on the building, and after the volunteers had left, some to service in Algeria and India, others returning to their jobs or studies, we missed them more than ever they would know, but the mark of their labours remained to remind us of their selfless endeavour – a wonderful summer of hard work, laughter and achievement – and this can be said of every successive team. They also added life to the village, inviting our friends there to an occasional campfire supper in exchange for hospitality offered, as well as weekly baths. They also received numerous gifts of fruit and vegetables as well as bread from a local baker. After they had left, the men continued with the building for that season, as long as our meagre funds and the weather allowed. They understandingly found other employment until we were in a position to go ahead again.

When Christmas came round it also brought with it considerable sorrow, with the parting in the physical sense of my mother in her 84th year, and also Florence Rowbotham, a life-long friend and the donor of the site. One was deeply thankful that they both reached the end of their journey peacefully. Although this is an experience shared by all at some time or another, there are no words to express adequately the feelings about such partings. Other friends had left us too, and while life will be considerably poorer without them, we shall think of them often with gladness and feel sure they will continue to support us from their celestial home.

Joy and sorrow, light and shade – this is the pattern of life, but in the end the light must surely prevail, for in the words of Tagore's assurance, 'Death is not extinguishing the light, but putting out the lamp because the dawn has come.'

20

The Walls Rise

During the following year, 1965, work on the Guildhouse continued to make good progress in spite of indifferent weather and exceedingly tight budgeting. IVS generously co-operated with us again by sending teams from the beginning of June until the end of September. Among the volunteers were several who had worked with us previously and we were delighted to renew old friendships and forge new ones. It was an enriching experience on all sides and the problems which arose were remarkably few, and related more often to external circumstances: like the difficulties involved in improvising scaffolding, which was always a headache. One idea was to draw the internal ceiling rafters through the walls to the outside of the building and use them as joists on which to place a temporary platform. When finished with they were pushed back and the holes in the walls made good again without a trace.

As the building developed we had to resort to old telegraph poles, standing them in weighted barrels. The ingenuity of the men knew no bounds and mercifully there were no accidents more serious than stepping on a nail in an old plank. A friend from the village who was employed by a firm of estate agents and auctioneers also supplied us with countless wooden bedsides, the residue from old wire mattresses. These were invaluable, especially for shoring purposes when a ten-inch layer of concrete had to be poured over the basement area to form both a ceiling and also a floor above.

Such an occasion could well have been a shot from a comic film as we all stood around with shovels, spades, rakes and every other tool and implement available, to await the arrival of the ready-mixed concrete. As soon as the mixer started to disgorge its porridge-like contents, we all dived in to spread the load as evenly as possible over the whole area. On one occasion the big end of the mixer suddenly gave out, roaring like a cannon, as it was pounding up the hill with the last load. There was nothing for it but to tip the slurried contents down the track – an ill wind, but it surely hardened the surface for future vehicles. I recall it was a brilliant moonlight night, which was fortunate for the men who were called out to deal with the crippled monster.

Meanwhile the appeal efforts were continuing, when time and space in the working programme allowed. It was a tough uphill struggle, but here and there a breakthrough in the form of an unexpected contribution added to the excitement for days; sometimes it was the exact sum required for a specific need. This happened when we had to resurface the original track with more hard core and chippings. Within a few days of receiving the estimate from the quarry the exact amount to cover the cost arrived in the form of a legacy about which I had previously been totally unaware.

Throughout the whole project whenever something (for instance in the way of materials) or someone was needed for a specific purpose, that need was fulfilled. I never had any doubt about solutions being found as each problem or requirement presented itself. With faith and patience all needs were eventually resolved, so much so that there were times when I was afraid to pray for an elephant in case one came walking up the hill! Needless to say, the serious side of this experience found expression in wonderment and profound thankfulness that the whole project was surely in God's safekeeping. And 'if God is for us who can be against us?'

1966 was a year of tremendously hard work. The early rains made conditions very difficult for the student volunteers who had gathered once again from many parts of the world to help with building operations. This year these were exceedingly demanding in the effort to reach roof level. However, by the end of the summer this had been achieved, with the exception of some final stone work on two gable ends, completed eventually by the stone-mason.

By October, Dave completed his share of the project. He had done a magnificent job over the previous years – especially in conjunction with the IVS teams, and we all appreciated so much the part he played in helping to get the Guildhouse out of the ground as well as his endless resourcefulness in dealing with the many heavy structural problems involved.

As the building slowly grew out of the ground, the experience of watching it unfold seemed very familiar. On recollection it was undoubtedly reminiscent of one I had as a child when the house in Wilmslow which my father designed came stage by stage into being. As the walls grew upon the foundations my sister Margot and I had the excitement of picnicking inside the rising shell, the dividing walls providing the demarcation lines of the various rooms. At a later stage we were also drawn into many of the discussions about the fitments and finishings, all of which had to be of the best quality available – a criterion which has remained with me ever since and is reflected in the construction of the Guildhouse.

21

The Symbol

Symbols representing a truth beyond their visual outward expression have always played a part in the more ritualistic religions of mankind. These have taken many forms; from the mandala to the crucifix. For us it seemed necessary to symbolise an ideal which would lead people beyond the personal, as represented by me, with whose life the conception and realisation of the Guildhouse had been so intimately involved. It seemed essential to formulate a symbol which would embody the ideals upon which the whole venture had been founded and by which it would continue to be motivated and sustained.

I had already been deeply impressed by the Benedictine ideal that work and prayer should be the twin aspirations for our whole endeavour, so it seemed appropriate that we should use the symbols for prayer and work, uniting them in a telling mandala. This would be placed in a strategic position where it could speak to us continually, and to all who were able to receive its message, bringing a sense of worship into our life and all our activity together.

So it was to be a cross, the symbol of worship, sacrifice, peace and resurrection, and a wheel, signifying all labour, the oneness of the family, the community of mankind, wholeness and eternity. Having established the symbol clearly, it was necessary to find someone who could translate it into wrought iron. Bill Martin, a local farmer, undertook to make it in his own forge, and I still recall those afternoons with a thrill, as every problem was gradually overcome. My friend was not a trained blacksmith, but he understood the inner significance of what was for him an unusual and difficult task. When the main structure was completed, I was allowed to do the final hammering to add a texture to the surface. I was reminded that one's life is refined by the fire of the Holy Spirit and beaten into shape, not of course by physical violence but by the hammering of God's voice upon the inner consciousness.

To return to the symbol itself, when it was finally placed in position at the end of the main hall, it was found to be just the right size, and as one casual visitor remarked, 'It certainly says something'. We all hope it will continue to speak on our behalf for a long time to come. Inspired by the symbol in iron, the mason decided to translate it also into stone. This we set above the main doorway leading on to the terrace, a tribute to the mason's craft and a sign of our fidelity to the Christian faith which has set its seal upon the whole venture.

For the work which now lay ahead and which we hoped to complete during the following year, a number of specialist tradesmen were needed – carpenters, plumbers, plasterers and electricians. Meanwhile Jim the mason, helped from time to time by his brother Jack, continued with the stonework, which also included the construction of the chimney stacks, in themselves a craftsman's job and a delight to the eye. Jim had also inspired a friend to carve the date stone which was inserted in the wall of the chimney at the rear of the Guildhouse, commemorating the year 1963 when building operations actually began.

By the spring of 1967, with all the outside stone work completed, our thoughts were turning to the construction of the roof, for which we were badly in need of funds. As ever, we were overjoyed when several charitable trusts responded to our appeal for help, so that the work could steadily move forward, and we could remain solvent.

22

The Cotswold Roof

The next task was to find a competent roofing specialist, but where was one to look for a craftsman with this very special skill? As always, help soon proved to be close at hand. On walking down the village street one morning I saw two men at the top of a ladder, re-covering with old Cotswold stone slates a roof they had previously dismantled. I watched for a time and, impressed by their expertise, waited for one of the men to come down the ladder, when I enquired for whom they were working. 'Cyril Miles, Moreton-in-Marsh,' was their reply. 'Do you think he would put a roof on for us,' I asked, and anxious to continue the dialogue I gave further particulars of our needs. The response to this information was, 'Well, there's no harm in asking him,' and having obtained all the necessary information, I contacted Cyril Miles. When I explained that the Guildhouse had been constructed largely with the help of student volunteers under the direction of two skilled men, the builder sounded a little apprehensive, but agreed to come and see what was really involved. At the appointed time he duly arrived and I shall never forget his first reaction as he stood in front of the roofless edifice.

'Well,' he said, 'something was taking care of you. I don't know how those youngsters got these walls so straight.'

He appeared really impressed by the quality of the stonework achieved by those unskilled volunteers. Consequently there was no difficulty in soliciting his co-operation over the roof. I had explained that we never undertook work for which we could not pay and that we were certainly in a position to go ahead with the first stages. He thoughtfully devised a plan whereby the project could be accomplished stage by stage as funds permitted. This involved erecting the roofing timbers, covering the whole with insulated roofing felt to keep out any inclement weather, and also to facilitate the continuation of much of the inside work. Finally the slates could be laid section by section.

We followed this procedure, and the roof was finally completed. Capped with an old stone ridge and with the addition of attractive lead flashings, it took on an air of true dignity equal to the quality of its superb craftsmanship. One carpenter remarked, 'One would feel it a sin to drive a crooked nail into this building.'

International Voluntary Service also came to our aid again that summer, supplying us with small teams of volunteers from the beginning of June until the end of August. They were engaged mainly in constructing a long retaining wall against the steep and subsiding bank at the rear of the building. This involved digging out tons of earth, replacing it with reinforcement and an inner wall before the actual stonework could begin. However, by the end of the summer the wall had been completed to the relief of us all.

Meanwhile the inside work was continuing, now that the roof was protecting the whole structure. We were fortunate in being able to call upon reliable sources which could supply all our needs in the way of materials. This information was made more readily available because of a list of suppliers I had compiled over the years, and to which I could resort when some special requirement was in question. Nearly all the materials, equipment and services were easily tracked down in this way.

Through this source five large oak beams for the main hall had been obtained from Messrs Groves of Milton-under-Wychwood. The original oaks had come from Blenheim Park and were lying in a field seasoning in the round. Having chosen five which seemed most suitable in terms of length and girth, the firm obligingly squared them up and arranged for them to be transported to Stanton. They were then dropped into the walls at ceiling height where beds had been prepared for them. This feat seemed to be achieved comparatively easily, with the aid of a tripod and tackle.

By great good fortune, many years earlier I had happened to be travelling by train from the North, when on passing through Stafford station my attention was drawn to an enormous timber yard bearing the name of Henry Venables. The name of the firm was duly noted and, when in the course of time we required timber for joists and rafters and eventually for floors, window-seats and sills, I contacted this firm. The result was a long and most helpful association, all our needs being negotiated most favourably by Alex Scott, their enthusiastic and exceedingly knowledgeable representative, who also succeeded in drawing his brother David into our orbit, encouraging him to make all the beautiful wooden latches for the doors, as well as several standard and table lamps.

The firm itself could not have been more helpful. They supplied us with lorryload after lorryload of superb English elm boards to meet all our requirements. They made all the doors, and the staircase which was a memorial gift from Ray Turner in memory of his wife Catherine, and which was designed by my nephew, Martin Wharmby. They also contributed a quantity of larch cladding for the kitchen, together with the gift of a lorryload of assorted oddments which Ray Shaw, a local craftsman, skilfully incorporated into the kitchen cupboards and other fitments, along with some sweet chestnut donated by Leslie Harris, which was also used under the staircase and for one of the floors.

During 1969 a great deal was achieved. Most of the plastering was done, and all the elm floors had been laid down and sanded, likewise the window seats, sills and most of the skirtings. Nearly all the doors had been hung and the old York paving stones brought to rest, so to speak, from a London pavement were laid very successfully in the great hall (as the stone-masons called it) and in the adjoining passages. The remaining stone wall and pillars were also completed. Most of the plastering was finished, also the wiring for electricity, and the plumbing installed for the central heating and domestic hot water supply.

During the summer of 1969 we also tackled the mammoth task of connecting our sewer to the mains below the Mount Inn. This was made a little less arduous by the willing co-operation of the local council who wanted to install some storm water pipes in the same trench. We reached an amicable arrangement whereby we would excavate a deep trench, lay our pipes and infill to a level where the Council could lay their storm water pipes. They then finished the infilling and completed the work by tarmacing the surface. This proved a very satisfactory effort for all concerned and the kind of exercise in co-operation which ought to be implemented more often to the participants' mutual benefit. It was also necessary at this juncture to build a pump-house near the gate to accommodate storage tanks from which a constant supply of water from the village main was to be carried into the building.

We also had help from another group of volunteers for a month, who did some decorating as well as clearing the site for further workshops to be erected when the main building was fully completed. A contribution from a local family had already been received towards the forge, to be installed eventually in the workshop still to be built. Another contribution towards the project was also received from an Indian friend, in memory of Mahatma Gandhi.

That year also we received the gift which put heart as well as efficiency into our kitchen, in the form of a large red oil-fired Aga, the generous gift of Donald and Mavis Rushbrooke, whose daughter Jill became one of our first residential volunteers. Through all the years since 'Pimpernel's' installation, it has remained one of our most faithful allies in providing hospitality for our many guests, and cheer throughout the bitter cold of winter.

Recalling these years, with all the endless activity and involvement with the outer concerns of day-to-day living, I am reminded again of the deep mystical experience which accompanied every challenge and its eventual overcoming. Nevertheless there were also times when one became utterly weary from the heat and toil of the day, and equally from the mental endeavour of holding all the disparate parts together. But work and prayer, rest and activity, must remain indivisible, for a divine thread links every experience together.

There were times when out of sheer weariness I experienced a void within, a nothingness which became vibrant again only after a considerable period of quiet meditation, sometimes aided by the rhythmical movement of my spinning wheel. This never failed to help me return to the task in hand renewed in spirit and physically refreshed. With this renewed committal came a sense of peace and inner communion, a strong assurance and a new energy. Then on to a new day. 'I wish you all success with the Guildhouse,' wrote one of that year's volunteers; 'I'm sure you will succeed, for so many people could not bear to see it defeated.'

23

The Dining Room Fireplace

My own initiation into building with stone, which had always fascinated me, began with working on a small area of the outside walls. My competence in this craft was considerably challenged however, when I decided to design and build the dining room fireplace. It was easy enough to visualise the task when completed, but the steps by which one arrived there were infinitely more difficult. This included sorting, measuring and dressing suitable stones which also involved the skill of breaking them when necessary in just the right place. It was heavy, arduous task, watched from time to time with obvious amusement by the stone mason working elsewhere in the building.

'I'm afraid I can't afford to employ you permanently,' he remarked jokingly, setting his rule against the walls as I proceeded, and after consulting his watch, recording the time I was taking in laying each course of stone.

When it was finished at last, apart from the plastering of the upper half, to be undertaken by the builder's men, I solicited his final approval.

'It's not bad,' he said, tongue in cheek, 'but you have omitted one thing. You ought to have inscribed across the lintel, 'Mary Made Me.' 'That would look more impressive in Greek or Latin,' I replied, laughingly.

Some time after the completion of the fireplace, I told this to Charles Tyson, then our honorary treasurer and a keen stone-carver, so much so that the impressive mediaeval dove-cot in his garden was regularly used as a workshop by an enthusiastic group of craftsmen in stone. It was there that Charles had carved an attractive stone lantern, already in use, to light the Guildhouse stairway. His response to my story was 'I am coming to do it', and sure enough when all the work on the interior had been completed, he arrived with the necessary tools and inscribed across the lintel in beautifully chiselled lettering 'ME FECIT MARIA'.

24

The Children

For me, contact with children, whether in East London or in the small hamlet of Laverton, was one of the most fulfilling experiences of my life. They were all equally engaging for their mischievous good humour as well as their often amazing insight. Life at the cottage was lived continually with the children, for they would drop in on their way to school or on their return journey home, to share with me their sometimes highly coloured news of the day, whether about their school, their families, the village or occasionally some wider startling issue involving the country at large. During my sojourn in their midst, they provided me with ample food for serious thought. Many of their sayings I recorded at the time, as they were too precious to be committed to oblivion, and some of their observations I shall always remember.

On one occasion the children were enraged on learning that Mr X had shut his dog up in a barn with the intention of shooting him. They came to my cottage in great distress.

'You must save Rover, Miss,' they pleaded. 'The boys can make him a kennel and we can feed him and take him for walks if you are away.'

'And after all,' considered one of them, 'What's the use of having a life if you are shot in the middle of it; it's better not to be born.'

'Mr X ought to come to spinning, then he'd learn to be kind to animals.'

On exploring this sad tale further, and to the great relief of us all, we happily discovered it to be an unfounded rumour.

I recall too the beginning of an Easter week when during the spinning session the question of war arose. The children must have sensed on this and on previous occasions my concern for peace-making. This perception prompted the following observation from one of them.

'Before the first Good Friday the people mocked Jesus, didn't they Miss?' 'Yes,' I replied, uncertain what was to follow. Then came the comment – 'And we'd be mocking you wouldn't we if we said there was going to be another war.'

During the years in Laverton I was just 'Miss' to the village children. This abbreviated form of address was not approved of by some of the adults who insisted that they should use my surname too – but the children were adamant and arrived at the cottage door to complain about this adult interference.

'Mrs X says we must call you Miss Osborn, but it isn't nearly as friendly, is it Miss.'

And Miss I remained for the duration of the Laverton days. The word still conjures up many warm memories. However, there was one memorable occasion when they did use my surname. It was Mothering Sunday and early in the morning the children arrived with a basket of violets and primroses, to which was attached the following note written in chalk, and which I still possess. 'To dear Miss Osborn, our second mother.' Throughout the years I have never had a more charming tribute, nor will those children, now grown up with families of their own, ever know the extent of that deep and abiding joy which they brought into my life. Their liveliness and enthusiasm for everything I did was a constant source of encouragement, and they in turn solicited my continuing interest in their concerns.

There were many humorous encounters too. Two of them arrived one day with the news that their father had acquired some dentures – a novel experience, for the children also.

'Dad can still take his teeth in and out. You see, they haven't growed in yet,' observed the younger sister, while the older one added, 'But there is always one thing that puzzles me Miss – whose were they?'

On the whole the village children tripped through their schooldays lightly enough, taking successes and failures, as well as admonitions, very much in their stride, but there were occasions which gave them cause for more serious concern. After a somewhat baffling scripture lesson, one of the older girls shared her dismay with the remarks. 'Oh, the Old Testament is so horrible, it almost makes me not believe in God.'

I tried to counter her obvious consternation by trying to explain that the Old Testament was really a history of the Jewish people's unfolding ideas of God, and about which the prophets had a clearer understanding. It was difficult to convince her of the essential message, but when I asked her what she felt about the New Testament she said, 'Oh, that is different – I can understand that.'

I wondered how I could present the tenets of my faith in a way that the children and young people could understand, during our Sunday gatherings at the cottage. These had been initiated by the children themselves in the absence of a Sunday School. This simple service was attended by Protestants and Catholics alike – the parents of the latter affirming that 'Children should be encouraged to pray in the home and the cottage is your home.'

A discussion group for teenagers also added another dimension. Each one in turn would suggest a problem, introducing it briefly, and this was then thrown open for general consideration. I was merely an observer who would throw in a comment occasionally with the rest. I also knew of the concerns expressed by some of their parents and could support their views discreetly, while encouraging these baffled teenagers to express their views quite freely. In this they felt secure, for it was an unwritten rule that nothing which was revealed through these discussions should be broadcast outside the group.

In retrospect this contact with children and young people, whether in the East End of London or in the countryside, was one of the most enriching experiences of my life. Their perception taught me more than they will ever know. Their loyalty and affection have remained with me always.

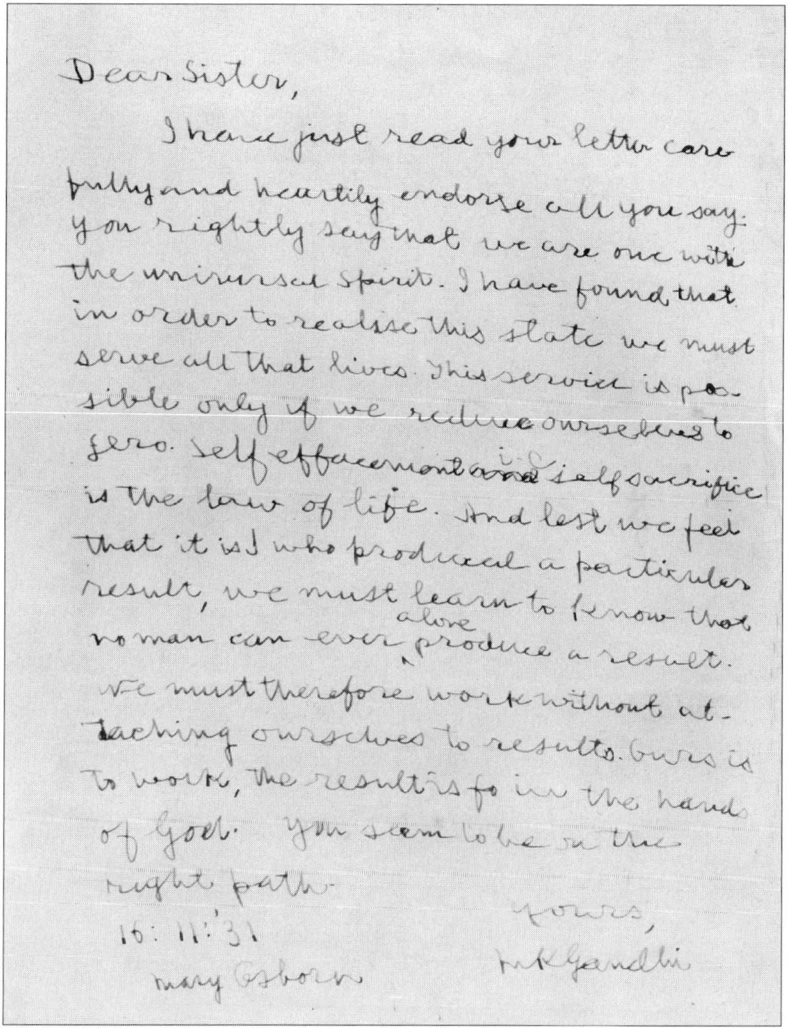

Dear Sister,

I have just read your letter carefully and heartily endorse all you say: You rightly say that we are one with the universal Spirit. I have found that in order to realise this state we must serve all that lives. This service is possible only if we reduce ourselves to zero. Self effacement ~~and~~ i.e. self sacrifice is the law of life. And lest we feel that it is I who produced a particular result, we must learn to know that no man can ever alone produce a result. We must therefore work without attaching ourselves to results. Ours is to work, the result is fo in the hands of God. You seem to be on the right path.

Yours,

16: 11: '31

M K Gandhi

Mary Osborn

Mahatma Gandhi, November 1931, in reply to Mary Osborn

The cottage at Laverton

Laverton children wearing their hand-spun woollen garments

Mary spinning on the steps of St Paul's Cathedral

Granpy Nash at Stanton

Jim Powell the stonemason, an Irish student volunteer and Mary building the south wall

Building in progress with Algerian students from the International Voluntary Service for Peace (*reproduced by kind permission of The Birmingham Post and Mail*)

Two views of the Guild Room. An elm chest is dedicated to the loving memory of Isabel Scoular, who saw the creation of the Guildhouse from the beginning.

Mary's fireplace, inscribed by Charles Tyson

The completed Guildhouse

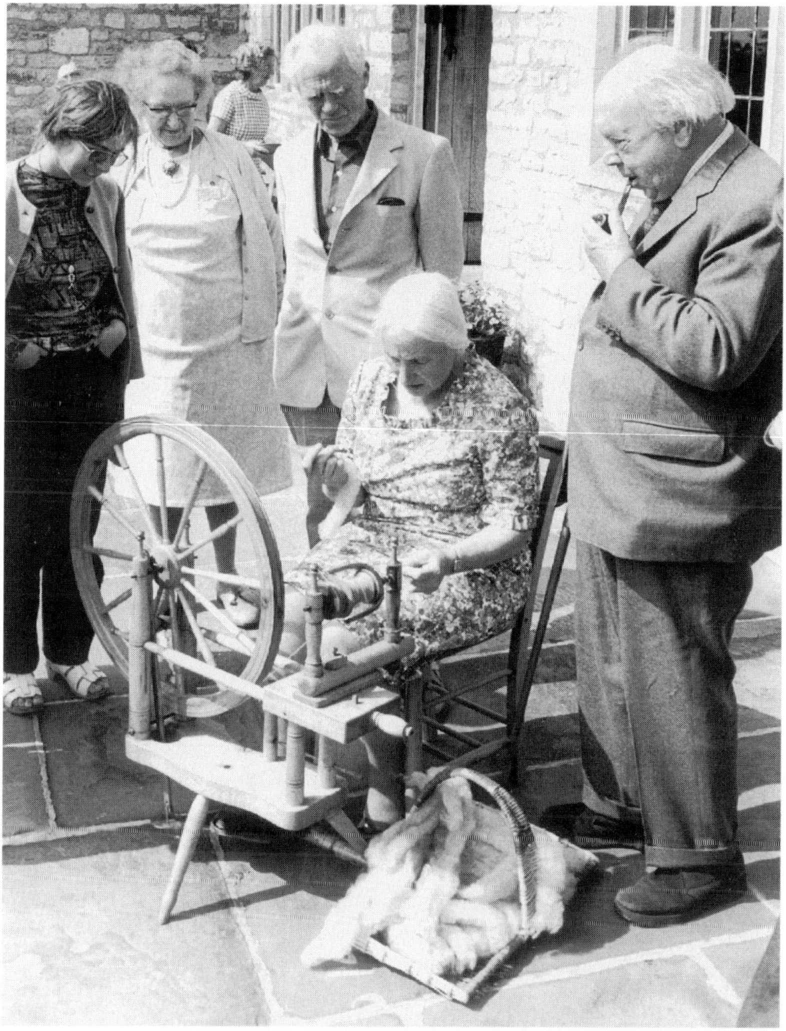

Opening Day, 1973, with distinguished guests and supporters: J B Priestley and Edward Barnsley (*reproduced by kind permission of The Journal Series, Evesham*)

An early woodwork group

Spinning beside the Guild Room inglenook fireplace

25

Integrity

As the walls rose and the stone mullions were inserted in preparation for the windows, one felt that more than a Cotswold building was being erected, but that each succeeding course of stone laid upon another with caring concern by the young volunteer builders was establishing a witness to a deeper spiritual dimension, culminating eventually in a sense of peace and wholeness. As a mentally handicapped girl enquired, on visiting us for the first time, 'Is this a house or a church?' It undoubtedly has the quality of both, hence my reply to her question – 'It is both' – and so it has remained.

When the walls had reached roof level, the roofing timbers were erected to carry the many thousand stone slates or 'slats'. Then the stone ridge was placed in position and the square stone chimneys together with all the attractive lead flashings completed the shell. One can truly say that the work was accomplished with the utmost care and concern by everyone involved. It now remained to match up the work on the interior with thc craftsmanship that had been applied to the exterior.

Some of the internal work had already been carried out, including the welcoming inglenook fireplace in what the volunteers called 'the great hall'. The large oak beams were already in position, but the elm sills, window seats, floorboards and the whole of the skirting, had to be placed in position, along with the doors with their traditional wooden latches and wrought iron fittings, some of this work being done after the plastering was completed. Quarry tiles were also laid in the kitchen and basement area, and the York stone slabs were laid in the main hall and on the terrace outside.

Kitchen fitments, using a variety of different woods, together with shelves and cupboards under the eaves, were all completed, along with the upper portion of the large chimney breast, providing several cupboards for clothes. The finished woodwork added a feeling of warmth and homeliness to the interior, while the white painted walls contributed significantly to a pleasant sense of spaciousness and serenity.

The experience of living in the Guildhouse has made me increasingly aware that the integrity of materials, along with the beauty of the basic structure, is never out of sight and in large measure is communicated through all the architectural details. Work and prayer expressed in our Guildhouse symbol have indeed been united in a common purpose and endeavour and will always remain an expression of the original vision of wholeness.

The development of one's potential, like the construction of a building, is concerned with the discovery of a fully integrated life. Consequently integrity must be extended towards every activity – the production and use of healthy food involving organic gardening and disciplined diet, the quality of craftsmanship in all its forms, expressed through the medium of natural materials – and involving nothing sham or pseudo. For myself, the weaving of cloth must also involve the handspinning and natural dyeing of the yarn. One is often criticised for this and dubbed a purist by enthusiasts for the use of commercial yarns and dyes. These also have their own integrity, to be used with honesty and not, I contend, as an attempt to imitate the handspun yarn and so mislead the uninitiated.

There is nothing more attractive or of greater beauty than the subtlety of handspun, vegetable-dyed and handwoven material when executed by the hand and eye of an experienced and dedicated spinner and weaver. This perfection achieved by a sensitive and dedicated craftsman can also be applied to other crafts – pottery, woodwork – including woodcarving – and calligraphy, to name but a few. It is recorded that Confucius was once asked what he considered to be the mark of a true craftsman and he replied, 'He is recognised by his honesty and integrity and by his ability to create something new with an experience which is old.'

The whole field of craftsmanship has been and still remains a matter of great contention in the world at large, resting so much upon that elusive quality termed taste – and in a world where so many goods are produced from materials designed to imitate something else. Ideally everything should be used as its essential nature honestly dictates. And now from the craftsmen's workshops to the market place. In relation to integrity there is one aspect of the ordering of our day-to-day existence which has always given me cause for great heart-searching. This is the whole question of finance, which in modern times is a far cry from the tenets of the Sermon on the Mount – like many other aspects of living. Over the years, there have been attempts to reform the whole financial system, but without any success. Also the present pattern of investments is so deeply entrenched in our national economy. Short of returning to a system of bartering – which has been attempted in a limited way from time to time – the present system is likely to remain for many years to come. One must also argue that there is a case, within limits, for the claims of its protagonists. However, a considered judgement about the true function of money is that it should be purely a means of exchange and not be regarded as a commodity to be bought and sold. This has prompted some would-be reformers to offer interest-free loans to people in need of the initial capital required for launching new projects in response to specific social needs.

The greatest problem arises when it becomes involved in activities related to chance, where large sums are extracted from the many and bestowed upon the few or even upon just one person. This seems to be a wholly un-Christian and therefore unethical activity. The means and the end should be in accord, but it will probably be many decades before the breakdown of the current systems forces the financiers to find a better way. In the meantime, one has to fall in with the least compromising of the options at present available and cultivate a certain detachment from money as such, at the same time making sure that it is being used for constructive and creative purposes.

26

Reverence for Life

One aspect of life which is often keenly debated is the whole question of vegetarianism, or the practice of non-violence in relation to the animal kingdom. This is based first and foremost on the humanitarian concept of reverence for all life, about which there is a steadily growing perception especially among the younger generation, who no longer consider it the prerogative of so-called cranks. The search for a more holistic approach to life relegates man to his true place in the vastness of creation – as one element, although an important one. Having the possibility of communion with God as well as an understanding of His precepts, man can only fully function within the orbit of all created life. He must take into account his true relationship to the animal and vegetable kingdoms, as well as to the whole cosmos. Symbolically, when man is troubled, so also is a star!

It is a fearful thought that creatures which man can never create are daily subjected to great cruelty, and many indeed to eventual slaughter, for which in many areas of the world there is no valid excuse. When American Indians kill for food, it is done with deep regret and is deemed a great sacrifice on the part of the animal whose death is surrounded by a solemn ritual. Killing is never perpetrated thoughtlessly for killing's sake. One also hopes for a time when such a so-called sacrificial necessity no longer claims any victims. Then in the words of the prophet, the animal kingdom will truly be at peace. In that day 'the wolf shall dwell with the lamb and the leopard shall lie down with the kid; and the calf and the young lion and the fatling together; and a little child shall lead them... they shall not hurt or destroy in all my holy mountain.' (Isaiah II : 6-9) Until that idyllic age, as prophesied by Isaiah, many people will never question the right to kill – often for so-called sport as well as out of the necessity for food – while other alternatives often remain unexplored.

As is obvious from existing archaeological remains, people of the earliest civilisations settled round the Mediterranean were horticulturists, that is grain and fruit eaters, and it has been declared one of man's most retrogressive steps when he became a meat-eater. Neither is man physically designed to deal with a carnivorous diet. Vegetarianism also excludes meat for health considerations. Meat often encourages an intake of too much protein, and is one of the most acid of foods, accelerating a number of rheumatic ailments as well as duodenal ulcers. To extract the maximum of nourishment, the bulk of one's whole food diet should be eaten raw. The brain, it is claimed, also remains much clearer. But the needs of the body in the last analysis largely depend on each individual's metabolism, so it is a matter of listening to the body and then following its dictates.

In relation to vegetarianism, compassion and diet are two of the most basic considerations; but there is also a third aspect which calls for immediate action. It has been stated by the World Health Organisation that there would be a greatly diminished world food problem if there were a universally accepted cessation of meat-eating. This is so wasteful of land, supporting comparatively few cattle to the acre. Such land could be utilised instead for the production of grain, pulses, fruit and vegetables. So from a purely practical point of view, this is a sensible consideration.

Underpinning the whole discussion is what might broadly be termed the spiritual dimension, which includes a feeling of relatedness which allots animals their rightful place in the scheme of things and at the same time shields them from unnecessary exploitation. Another aspect which is ethically of the greatest importance involves allowing others to kill animals on our behalf. Most people would never eat meat if they were called upon to do this degrading task themselves.

Apart from breaking the chain of divinity which links the whole of creation together, and contrary to the often stated contention that God created animals for the use of man, Genesis specifically states: 'God said – Behold I have given you every herb bearing seed, which is upon the face of all the earth, and every tree, in the which is the fruit of a tree yielding seed; to you it shall be for meat.'

While pondering on these thoughts it is also well to recall that a simple vegetarian meal, taken three times daily, and consisting of goat's milk, two oranges, with a few grapes or dried raisins, gave Mahatma Gandhi sufficient energy for each day's arduous tasks. If he could do it, so can we, if accompanied by prayer and vigilance. For 'man does not live by bread alone', but by his moral precepts borne out of his spirituality and reverence for all life.

27

The Doors Open

We made it at last, or nearly so, for at the beginning of July 1970, I moved into the Guildhouse permanently, just seven years after laying the foundations. Eva Hall, a Swedish volunteer of four years previously, came back again that summer to help us. They were unforgettable days of sunshine and hard work and Eva was as happy as I to see this dream of so many years nearing fulfilment – her home in England, as she called it.

The move from Laverton was less joyful in that it seemed interminable. The cottage which I had rented for so long was becoming daily more derelict. I had tried to keep it habitable over the years, but it needed many major repairs, and by the time I left the thatch was leaking badly in many places, so in spite of many happy memories I did not regret leaving it. That moment was the end of a chapter and I could look forward to the next with a feeling of heightened anticipation, for it would offer a far wider horizon with greater possibilities for usefulness. As to the move itself, I was deeply indebted to some local friends who voluntarily transported all my goods and chattels – load after load of them – from Laverton to the Guildhouse.

While the Guildhouse was at that time more or less habitable, a good deal of finishing remained. This was also exciting as one by one the essential jobs were finally completed and we could put them behind us. These included the installation of the central heating system, the final stages of the electrical and plumbing work, and the completion of the whole basement area, comprising the heating chamber, utility room and accommodation for the pottery, all of which had still to be wired, plastered and the floors tiled.

Ray Shaw, a local craftsman who was working on the kitchen fitments at that time, also proved to be a most excellent baker. Getting up at 4.30 a.m. he would often bake bread in time for breakfast, much to the surprise and delight of our many visitors. He also took over the vegetable garden which we were wresting, little by little, out of a bramble thicket, and which in spite of fierce competition from the rabbits produced some excellent crops.

There was plenty of work for the many volunteers, young and old, who came to our aid again that memorable summer. Some arranged weekend workcamps, others arrived daily, armed with a favourite tool to augment our meagre equipment. The whole site was in constant need of clearing, fences had to be repaired, flower gardens laid out, and the roadway, being only of a loosely-knit kind of texture, had once again to be patched up. It was not until many years later that a more permanent surface could be afforded.

As far as the building itself was concerned, a high standard of craftsmanship in all the work undertaken had been maintained throughout with a deep sense of integrity. It was a delight to behold and a joy to live in, largely due to the love and concern of all who had participated in its creation. Expressions of the emotions felt were often given voice to by the volunteers themselves. One such comment was received in a letter from a Belgian volunteer who had been exhilarated by the experience of taking part in the construction of a building to be used in such a creative way. Having said how often his thoughts were with us and how he hoped that many people, old and young, rich and poor, would be able to come to the Guildhouse when completed, he continued, 'for there they will find friendship for the heart, beauty for the eyes and peace for body and soul.'

One such need arose shortly after I had taken up residence. A local couple, with their two sons, had recently sold their small-holding, believing in good faith that the completion of the purchase of another house was imminent. As so often happens, another buyer made a slightly higher bid and the would-be purchasers were left high and dry; in this particular case, without even being given the chance to increase their offer. The situation put them in a real predicament, so we offered them accommodation at the Guildhouse until, some weeks later, they were able to acquire another property. It was a mutually helpful arrangement, and an experience of true sharing. When the time came for them to leave, we sadly missed our ready-made family.

Meanwhile the final fittings were being placed in position – the wall lights, standard lamps and the remaining lamp-shades for all the rooms – spun and woven or wound from handspun yarn, adding their own touch of originality, as well as being extremely practical. Handspun and woven rugs enhanced the floors, and bedcovers spun from Jacob's fleece and others dyed scarlet by the local children for the children's room, were woven by the village girls and other teenage visitors. To add to the homeliness of the main hall, stone seats were built inside the huge inglenook fireplace, the log fire being contained in a wrought-iron basket, with protective dogs and a canopy above – adding a warm heart to the friendly scene, especially in mid-winter. To complete the radiance, a stone lantern, carved by our former treasurer, Charles Tyson, and inserted into the wall above, illumined the whole staircase with a pleasantly subdued light.

We were very fortunate that the house attracted a number of gifts in the way of furniture, a bolt of white cotton sheeting, kitchen equipment, garden tools and two wheelbarrows, one of which was contributed by the village children from their own savings. The firm which supplied it was so impressed that they gave us a second barrow. The boys also donated some woodworking tools, from the sale of wooden toys which they had made, and which they bought when we arranged a special shopping expedition to Gloucester. This was a beginning, and over the years since, everything we have needed has been supplied eventually, whether materials, equipment, volunteers or the draftsmen required to complete a specific job.

As time went on, each year brought its own special joys and benefits. We were conscious of being wayfarers along the path upon which our feet were firmly set. The overriding concern was to set our sights in the right direction in the hope that others would travel with us.

28

The Dream Comes True

By 1971 it was possible to re-establish some of the local craft groups under the Guildhouse roof. Meanwhile there were still a number of practical jobs to be accomplished, including the resurfacing of the track up to the Guildhouse, to convert it into a really accessible roadway. Members of a local charitable organisation offered to do the work with the help of volunteers, but on further consideration decided it was too difficult and heavy a project for them to undertake. In the end, two of Granpy's grandsons, Sem and Vic Wilcox, generously offered to do the work without payment. The offer was so spontaneous.

'If you'll find the materials, we'll put them down,' they said, and sure enough they straightway set about this heavy task with good humour and the kind of expertise necessary, for their daily work was with the county council. They laboured many evenings, Saturday mornings and during part of their holiday and did a wonderful job, for which many people – especially the aged and handicapped – were profoundly thankful.

1971 also saw the completion of many other major jobs, including all the plumbing and electrical work, the installation of the entire central heating system and the fitting of the main oak doors with their craftsman-made wrought-iron fittings. Outside, the remainder of the York stone slabs from a London pavement were laid over several areas, adding to the feeling of age which seemed to characterise the whole building. The builders also completed the ground floor rooms at the lowest level, which included the heating chamber, the pottery, and accommodation for domestic washing, as well as for dyeing the hand-spun yarn.

Because of the use of old stone for the exterior, many people have been misled into thinking that the Guildhouse is a genuinely old building which has been reconditioned and modernised. 'Was it in a very bad condition when you took it over?' many people have enquired. When informed that nothing was on the site except brambles and fallen trees, and that we had started the building from scratch, they could hardly believe it.

With the completion of the building work in sight, family and friends showered upon us all kinds of gifts in he way of household and office equipment. These also included a potter's wheel, an electric kiln, a garden roller, a mower and other essential tools. Little by little the entire pattern was taking shape, while the ideals upon which the whole venture was founded were also becoming more manifest. One of the rooms was equipped for woodwork, so that men and boys, including fathers and sons, were provided with the facilities for working at a craft together, under the instruction of two experienced volunteer tutors, Arthur Adams and Eric Dinwiddie, who later became our enthusiastic Treasurer and a most supportive friend of the Guildhouse in every way. This accommodation was a temporary arrangement until a purpose-built workshop with space for a forge and storage space above could be erected some time later.

Meanwhile, seven small weekly groups for spinning and weaving were re-established, with two special groups for mothers and children. This proved very successful until, in later years, the exodus of young families into the towns accelerated, and sadly the child population decreased; the smaller village schools were closed and many a school bell has hung idly in its belfry ever since. With the closing of our village school, the life of the village became noticeably more sober; the older people especially missed the children's cheery voices as well as their ever-ready willingness to run the occasional errand, for a small reward or none.

The original vision for the Guildhouse had included provision for accommodating groups of children residentially, whether deprived, handicapped, or physically fit; any children, in fact, who could benefit from a creative experience, in the way of sharing life generally or working together on a craft or with music. Two of the upstairs rooms were consequently turned into small dormitories, equipped with bunk beds with handspun covers, dyed and woven, some by the children themselves, while other necessary facilities were also provided.

The summer of 1972 also found us preparing the foundations for the remaining workshops, designed voluntarily by a nephew, Philip Wharmby, and promising the space we so badly needed for woodwork and ironwork, as well as for the storage of materials and seldom-used equipment. The work was again aided by local volunteers as well as members of International Voluntary Service. All that remained to be done was to raise the funds for the completion of the workshops. This included the cost of some extra stone, a number of old stone slates and ridge for the roof, the windows and doors.

In a project such as this the main tasks which have to be done are obvious, but it is the seemingly countless small jobs that prove to be so time-consuming. We were therefore exceedingly grateful to Bill Wootten, who addressed this need so conscientiously over a long period; also to his wife Christine and their son Anthony who have likewise supported us very generously.

Our needs over the years were also the concern of members of my own family. My sister Margot and her husband Ewart Wharmby (who was one of our first Trustees) together with their family have been most supportive of the whole venture, each contributing their measure of creative skills. This can also be said of the Trustees, as well as the Friends of the Guildhouse, now a separate registered charity.

Meanwhile, the development of the garden was largely due to the really hard work, imagination and generosity of Amanda Pallant, who over the years donated and planted a well-chosen variety of trees, flowering shrubs, moss roses and rock plants, to which have been added a number of special items, including a variety of soft fruit bushes. In due course, everything was provided for by local efforts, generous gifts from friends, and also by the encouraging support of several Charitable Trusts.

At this stage our thoughts were turning towards the dedication of the main building, as interest in the whole venture was spreading far and wide. Some of this interest might have been lost had we delayed the official opening until the workshops were completed.

29

The Dedication

It was a glorious Whit Sunday in 1973 when we finally dedicated the building to the greater glory of God and to the service of our fellow men, women and children. It was the inspiration of our minds, the intention of our hearts and the labour of our hands, all God's gifts, which we were offering to Him again. The whole service centred around the message of Pentecost and the gift of the Holy Spirit which had brought to fruition so many facets of our endeavour and seemed to permeate the very walls so lovingly constructed. There was also a deep sense that we were giving the tired old barns, which had yielded the beautiful vibrant honey-coloured stone, a true resurrection.

The many friends present had themselves caught the vision and added their strength and inspiration to its fulfilment. We were merely channels for the work which God needed to be done in this very small corner of his world. We were also celebrating once again the coming of the Holy Spirit like a 'celestial gale' into the hearts and minds of the waiting disciples, linking in our thoughts their experience with our own.

Through the long, struggling history of man, one sees the movement of the Holy Spirit at work – sometimes gentle, at other times more forceful – through the lives of saints, seers, prophets, musicians, poets, artists and also through the work of gifted craftsmen. It has inspired countless men and women of every age – to be good neighbours, grateful friends and responsible parents – all who have been true to the noblest concept of life of which they were capable.

Significantly for us, we find its fullest expression in the West in the life of Christ – a life lived fearlessly in the world, but motivated through prayer and contemplation by the other-worldliness of God's Spirit. Perhaps we can best understand this by Christ's own interpretation of love. He left us two great commandments: to love God, involving the seeking and the doing of His will; and to love each other, meaning the identification of oneself with another person's need, whether of heating, comfort or forgiveness. This power of love was set free at Pentecost with tremendous energy, and inspired the deep sense of communion with Christ and with each other, so evident in the life of the early Church. We read that 'all that believed were together' and had all things in common. They sold their possessions and goods and parted them to all men as every man had need. How far we have travelled from that ideal!

How far is all this relevant to the building of the Guildhouse? Personally, I believe that all my life has been directed towards this event. Looking back over the past years it seems that everything that has happened has moved in this direction. There were long years of preparation, not without signposts along the way, for many people inspired my early endeavours. I owe a great deal to Christian ideals and to the encouragement of Mahatma Gandhi, one of the world's great Christians in everything but name. I also owe more than I can say to my own family and friends from many parts of the world as well as in Bow, who have shared these ideals and had given the support of their friendship, prayers and practical help so unsparingly, as well as sharing their many creative gifts. And there were always the children, who in the struggling Laverton days filled my life with joy and encouragement. Happily there are still children whose company is always a delight – some of them children of the one-time Laverton children!

If in the days which lie ahead, we can recapture through work and prayer the spiritual energy of Pentecost, which made the life of the early Christians so vital and contagious, and if through the Guildhouse someone finds a little more peace, courage, compassion, joy, hope or inspiration for living, then God will not have entrusted to us in vain this small vehicle of His Creative Spirit.

These thoughts, together with the hymns, readings and the music of Beethoven and Busoni flooded our minds as the events of that memorable week flowed on, while hundreds of friends and well-wishers came to share the joy of a dream come true. Many also brought gifts. One of these which moved me deeply was a handmade spinning chair, with our symbol carved on its high back, and bearing also my initials and those of the donor, James Ashman, who was the senior building inspector for this area. He had been my constant encourager, guide, counsellor and friend from the moment we lifted the first clod of earth in preparation for the foundations. Subsequently the same friend made for us a beautiful lectern.

It was a warmly giving week in every way, with many old friendships renewed, culminating in an unforgettable recital of music. As the strains of Haydn, Beethoven, Mozart and Bach floated through the open doorway to the hills beyond, where sheep and cattle grazed peacefully, it had, as one friend expressed it later, a heavenly quality.

30

After the Dedication

During the warm sunny days immediately following the Dedication hundreds of friends arrived to share our happiness at the completion of the main building. Some of our supporters, prominent patrons in the fields of craftsmanship, the arts and the sciences, were also eager to be present and willing to discuss with interested visitors their common concerns – an opportunity warmly appreciated on all sides. Among those who generously contributed to the occasion in this way were J. B. Priestley, Edward Barnsley, Mary Adams, Dr Barbara Latto and Professor C. A. Coulson.

Music, which has played a vital role in the life of the Guildhouse, added a fitting finale to the week's festivities. The occasion was recorded at the time by the Editor of the Evesham Journal, C. W. Clark. He wrote: 'It was a week of open house which began with a Sunday dedication, contained a variety of daily events and ended idyllically on Saturday with a musical evening that will not be forgotten by any who had the pleasure of being there. Lina Vincent (violin), June Everett (violin), Charles Tyson (viola) and Phillis Travis (cello) played a Haydn quartet – the one with the most famous of minuets; Penelope Foster (cello) and Roma Foster (piano) played Beethoven's Sonata in G Minor Op. 5 No. 2, and Elizabeth Baker (soprano) accompanied by Raymond Cantwell (piano) sang Mozart's Alleluia and several other songs. There was also some unaccompanied Bach.'

The Guildhouse dream was a reality, a superb example of building in natural stone. In all its finishings and furnishings it quietly asserted the dignity of excellence in craftsmanship. This was the finest work of practised hands. Even when a dream is realised, however, one must always go back to work, at the same time continuing to dream the right dreams.

31

Campden Rugs

A special kind of joy is experienced when a group of people come together spontaneously to work on a communal project. It so happened that when one of our members, Margaret Wright, heard that the Lady Chapel of Chipping Campden Church was badly in need of new floor coverings, the idea came to her that these could be hand-spun and woven at the Guildhouse, if a sufficient number of people would be prepared to work on them.

When the news was spread around it was greeted with great enthusiasm and offers of help from twenty people. As Chipping Campden is one of the Cotswold wool churches it seemed fitting that fleece from the original Cotswold breed of sheep should be used, along with some Suffolk wool to give it 'body', mixed also with a little Shetland moorit (a light red-brown colour) to turn the whole into a soft oatmeal shade, picking up the colour of the natural stone walls of the interior.

Most of the groups were involved with the preparation of the wool. This was first teased by pulling the fleece gently apart without tearing the fibres. Then a mixture of the white and moorit fleece was brushed between two wire-toothed carders and finally made into rolags or rolls of wool ready for spinning into thick yarn by two of the spinners. The preparation of the wool needed many hands; spinning the thick yarn was done quite quickly. Two spools of the thick yarn were then plied together to give it extra strength, the wool was then skeined, and washed ready for weaving.

Meanwhile I was preparing the long flax warp, but by a different process. A bunch of flax was tied to the waist and spread out fan-wise, little by little, from left to right and right to left alternately to form a kind of apron. This was then rolled on to the distaff, tied in position at the top and bound lightly on to the staff with a coloured braid. The strands of flax for the required thickness were then continuously pulled down from the distaff with the left hand which must always be kept dry, then guided on to the spinning wheel while it is in motion, with the right hand which must be continually moistened by dipping into a bowl of water. Two spools or bobbins of flax were then plied together, skeined and washed, when they were ready for use. This process was repeated until the required length for the long warp, across which the wool was to be woven, had been completed. The warp was then transferred to the loom from the warping board, on which the required number of threads to a specified length and width had been prepared for the final process of weaving.

When the work was finished and the floor covering laid in position, a little ceremony of thanksgiving added a special blessing to all who had taken part in the final achievement.

32

The Spiritual Dimension

As far as the spiritual dimension of life is concerned, our monthly Day Retreat meets a need among people of all ages and from many walks of life. Its membership is fully ecumenical, covering the whole spectrum of Christian denominational loyalties; it tries at the same time to understand the essence of other world faiths, not espousing uniformity, but diversity within an overshadowing unity. A central theme is illuminated by readings and prayers interspersed with periods of silent meditation. Sometimes a discussion on the chosen theme follows and there is always time set aside for intercession – perhaps the most unifying and healing of all our corporate spiritual experiences and an undeniable channel for the action of the Holy Spirit. The whole day provides opportunities for those who share it to find their own centre – the still point within themselves where tumult ceases.

At the last supper Jesus shared the bread and wine with His disciples as symbols of His sacrifice for all mankind and for their inner spiritual renewal after His resurrection. And so they are for us today as we feed on His life and teaching in our hearts 'by faith and with thanksgiving'. Most of life is exceedingly mysterious, shrouded by the 'cloud of unknowing', but from time to time there is a fleeting break in the cloud through which the love of God and the nearness of Christ illumines for a brief moment our lives. Then the here and the hereafter are linked by the light of immortality, before the contemplation of which one can only stand in awe and reverence.

33

A Christmas Celebration

One of the most precious times of the year is our Christmas carol service, with an accompaniment of music arranged by a number of talented musicians, including singers, pianists, flautists, violinists, guitarists and on one occasion by a harpist. Each year a different musical offering supports the well-loved carols.

Beforehand the hall is decorated with holly, ivy and branches of evergreen fir, and on every window sill shines a tall, red candle. There is also a traditional mistletoe bough hanging from one of the great oak beams which support the ceiling, upon which a host of shadows flicker. A tall tree stands in a far corner reaching from floor to ceiling and bearing a host of small red candles, while standing on the chest beneath the Guildhouse symbol is a lighted crib.

The silence gives way to the sound of singing, the Christmas story unfolds as the carols are sung, the lessons read and the prayers offered. We hear of the love and humility of God, sending his Son to serve and heal the poorest of men, having taken upon Himself the flesh of our humanity and the mantle of poverty. Yet like the star which guided the wise men into His presence, the light of His glory illumines the whole world. Then follows the most moving moment in the service when the candles on the tree, representing so many human needs are lit, one by one, as absent families, friends, the sick and the sorrowing are remembered by all who care. When all the candles are alight, the service ends with the singing of that immortal carol, 'Silent Night, Holy Night', then the final blessing of peace is given. As they leave, each family is given a candle to light in their own homes on Christmas Day, so linking all those who have shared in the service with Christ and with each other.

34

The Life Within

Since the completion of the main building in 1973, the Guildhouse has slowly taken on a life of its own and has been variously described by visiting friends as a rural retreat, a shrine against materialism, and a modern expression of monasticism. Although the last two claims are somewhat extravagant, the building does have an air of timelessness and serenity. This is due no doubt to all the care involved in its creation, as well as the beauty of the surrounding countryside, with the gentle hills close by and the long vista beyond to the far-away river Severn and the Black Mountains on the Welsh border. In the foreground the morning sunlight often glints upon the majestic towers of Tewkesbury Abbey.

It is within this context that one sought to make the Guildhouse, as a handicapped girl touchingly described it in our visitors' book, 'God's home for all'. We were also anxious to encourage fathers and sons, mothers and daughters, to experience working on a craft together, and in the early days before so many young families moved out of the villages, this was possible. A number of volunteer tutors also began to offer their services, so that each craft group was led by an experienced teacher, while many gifted musicians arranged a series of concerts, thus sharing their gifts with an ever-expanding and appreciative audience.

We turned our thoughts towards inviting groups of children to stay with us. The year before the official opening, the first of these comprised twenty-six children from a school in the Bristol area. They were accompanied by two members of the staff, one of whom I had known since she was a child. Although we had purchased in readiness a number of bunk beds beautifully made in Yugoslavia, together with some of the bedcovers spun, dyed and woven by the village children, there were not enough to accommodate them all. Consequently we had to resort to filling much of the extra floor space with spare mattresses and air-filled lilos.

The first night was quite hilarious, for the children were so excited that they spent most of it visiting each other in the various rooms. Although they soon settled down and were exceedingly well-behaved and appreciative we realised that the group was too big and that in future we should arrange for them to come in two groups. This would mean that they could all eat together as a family in our somewhat limited dining room, instead of at long tables in the large hall. The children's energy and our enthusiasm knew no bounds and even today, so many years later, my heart warms to the remembrance of that experience. Their time was spent in spinning, modelling, exploring the countryside, making records of their impressions, and finally organising a concert with recorders and guitars for the benefit of their parents, who came to collect them all on the final day.

Letters received afterwards conveyed the pleasure of these experiences for as one girl wrote, 'Whenever I'm upset (which is fairly often) I think of the Guildhouse and it cheers me up'. Needless to say, many such visits have followed since 1973, when all the activities became more fully operative, meeting a number of specific needs. One thirteen-year-old girl, after her fourth visit, wrote 'I couldn't live without it', while the following moving tribute came from a fourteen-year-old: 'Thank you so very much for such a wonderful time. It was a most beautiful experience and to spin again was a marvellous sensation. You have no idea how fond I am of the Guildhouse. It is in such lovely surroundings and it is quite fantastic to wake up to the wooden floors and simplicity and to be able to gaze quite freely at the spectacular view and sunsets. I really can't thank you enough for giving me so much help and providing me with a most perfect holiday.'

On one occasion my niece, Alison Edwards, brought a group of the children she was currently teaching to the Guildhouse for a craft holiday. This experience prompted

Jo Carter, a ten-year-old girl, to write the following succinct verse entitled 'My thoughts of the Guildhouse'.

Grand but cosy
Stone but warm
New but old
Away but home.

Every successive summer when we had a break from the local groups, a number of weeks were set aside to meet the residential needs of many children, some of whom were seriously handicapped physically: like the group of Vietnamese orphan girls suffering badly from the after-effects of polio, or the numerous groups of older handicapped boys and girls from a rehabilitation centre, who stayed with us many times and gave us so much more by their courage and cheerfulness than we could ever give to them. This was also true of the Tibetan girls from a farm school in Wales who visited us on several occasions. And what joy they all gained by their simple achievements in drawing, spinning and weaving – the girls making attractive shoulder bags for their own use, the boys doing woodwork: on one occasion this was a sturdy and well-designed plant stand as a gift for their centre, made co-operatively.

Over the years our puppet theatre has also given great pleasure to many children, some as performers entering imaginatively into the various characters, while others along with their parents provided an audience. The majority of puppets were made by Elsbeth Weyel, a gifted German friend. The rest we made ourselves, also writing the scripts, as well as constructing the scenery and arranging the lighting effects and the music. Sadly, such plays as 'St Nicholas' and a fantasy entitled 'Christmas Comes to the Martins' seem to become less and less appealing with the present-day advent of more sophisticated entertainment, but one hopes that the magic of make-believe will still hold a fascination for some.

We had other happy experiences when we discovered that a number of children from the surrounding area were playing a great variety of instruments – trumpets, flutes, violins, recorders, guitars, a cello, harp, lyre, oboe, French horn and of course the piano. So on several occasions we invited them to come together to play to each other. These gatherings inspired one friend to provide us with a splendid selection of percussion instruments for which a fifteen-year-old boy, Tobias Alington, composed a percussion symphony with piano and trumpet accompaniment. In recent years this musical activity, with the support of Jean Alington and Ann Verney, has developed into a more formal children's concert, giving generous support to the Save the Children Fund and the Speedwell Trust for Music Therapy.

While all these and other activities were developing, the day-by-day work of maintaining the orderliness and smooth functioning of the domestic scene was being cared for by a series of voluntary helpers who arrived from time to time for longer or shorter periods. Of all these Martha Frei remained for the longest time.

Martha came to us in 1974, having arrived in England from Switzerland many years previously to help a German refugee family. After further work experience, she was introduced to the Guildhouse and stayed with us for eighteen years. During that time she made with complete dedication an invaluable contribution of the day-to-day running of the Guildhouse, and will long be remembered with much love. She now lives in her retirement with friends in Malvern.

35

The Workshop Rises

By 1976 plans for the remaining building to house all our wood-working facilities were coming to fruition. Work on the foundations had been completed when the oversite was prepared by a group of young student volunteers from the Cheltenham area. The building itself was then erected by Cyril Miles, the same local builder who had previously given us so much help. It is a simple functional building, designed by my nephew, Philip Wharmby, with workshops below for which an elm floor was generously donated by a number of firms connected with the Home Timber Merchants' Association. Above the workshop is a spacious loft providing ample storage space, while the exterior of the building has a wrought-iron balustrade and handrail made, along with a handsome lamp which lights the whole area at night, by a young craftsman friend, Anthony Wootton.

Over the years the workshop has been fully equipped with all the necessary tools, lathes and other simple machinery (much of it generously donated) to facilitate a number of woodworking crafts – turning, carving, furniture and toy-making. These have been added to by the very generous bequest of the contents of his personal workshop by the late Edward Purbrook. More recently we received a further welcome gift of tools which were left to us by Jim Ashman who, as mentioned elsewhere, was an enthusiastic supporter since the early days. Excellent tutors offer their services to an ever-increasing group of men and women who respond enthusiastically to the help given for improving their skills and their sense of design. Much of their work along with that of the spinners, weavers, calligraphers, potters and woodworkers would undoubtedly have earned the approval of St Benedict himself, whose ideal – *Laborare est Orare* – we have tried to embody in all the work of our hands.

As time went on a more efficient adjustable lighting arrangement was found to be necessary. This was very generously donated by my nephew David Wharmby and his wife Lindsey. Other contributions included the two main oak doors for which the wrought iron fittings were designed by Alan Knight, made by Ken Woolly and donated by my sister Margot and myself in memory of a dearly loved mother, herself a true home-maker.

It is always sad when friends with whom we have shared so many common interests and concerns leave this life and it is not always easy to decide upon a fitting memorial. Since Eric Dinwiddie, our former treasurer, was a very competent and enthusiastic gardener, it seemed appropriate to plant a number of fruit trees in his memory, hoping that they will endow the site with their bounty for a long time to come. This is also our hope for the grape vine and the pergola which supports it in memory of Ken Powell, a former trustee and an exceedingly competent woodwork tutor. Here too I must remember Ben Wright who when he retired brought his training as a physician to bear so diligently upon the constant care of all our tools and equipment.

36

Iona of the Dove

During the building of the Guildhouse from 1963-1973 and for several years afterwards, it was impossible to get away for what could be termed a holiday. A very short break of a day or two with family or friends was the most that could be fitted into a demanding schedule. However, the unexpected arrival of Jan who found her way to the Guildhouse during the summer of 1976, made it possible to embark on one of several hoped-for pilgrimages – the first to Iona.

Jan, an American, had felt a strong urge to come to England, where she felt she had work to do. By various contacts she was led to our door and remained here for the next ten years between occasional visits to America. On Jan's arrival we were badly in need of extra help and her interests and training as a gifted hand-weaver fitted naturally into the pattern of our life and work.

The spring after Jan's arrival our thoughts turned towards a possible visit to Iona. We set off just before Easter to find this holy isle, intent on spending the whole of the festival there. It was a long journey from Gloucestershire, even allowing for an overnight break, and we arrived on the second day too late to catch the last ferry from Mull to Iona. Fortunately accommodation for the night was not difficult to find. And early the next morning we made the crossing and found our way to the place where we had arranged to stay, as the Abbey itself was completely taken up with young people from the mainland.

We learned that when St Columba landed on Iona on May 12th, 563, he found the people steeped in mysticism, with a deep love of music and the arts. From that time onwards the island became a centre of intense missionary activity, converting Scotland to Christianity, spreading its influence throughout England and later to the continent of

Europe – the greater part of Germany and Switzerland and also a part of Italy. The height of its power was in the seventh century when Iona was undoubtedly a seat of learning, with a widely renowned theological school. It is believed that the convent at Erfurt, which eventually produced Martin Luther, was also a Celtic foundation.

The early Celtic Church mainly centred around monasticism, and the western seaboard abounds in relics of monastic cells and Celtic Chapels, many of their names incorporating those of former Celtic saints. However, the long and chequered history of the island was not always bathed in sweetness and light. This burial place of kings, princes, chiefs and ecclesiastics suffered also a tortuous history of invasions and massacres from marauding tribes from the north. But after each phase of destruction from without, there arose a spiritual resurrection from within.

During the early part of the 13th century, a Benedictine monastery was established on Iona, also a community of nuns of the same order, the ruins of whose nunnery still exist. This has remained for me of very great interest, since after my visit to Iona a friend, Yvonne King, gave me an historic silver cross, of authentic Celtic design, which was reputed to have belonged to a Prioress or Abbess of the same Benedictine nunnery. One cannot be certain about its past history, but it has been confirmed that it is the appropriate size for an Abbess's cross and it has the word 'Iona' embossed on the back. Also the edges of the cross on the underside are worn, as if continually chafed by a rough habit. How one wishes it could speak!

In retrospect it is difficult to convey the depth of emotion we felt on setting foot on such sacred soil, knowing how many centuries of history lay hidden beneath the now silent greensward and gently rising hills. Nor can I ever forget the first sight of the resurrected Abbey church, with the adjoining buildings standing quietly radiant at dawn, and in the evening of that first day bathed in the roseate light of the setting sun, a sure testimony to man's continuing faith in the everlasting power of God.

Memory also revives the knowledge of the ongoing spiritual life within the Abbey's majestic walls. Here those who enter feel completely at home and safe for a time from all the storms of life. It was a privilege during the days which followed to take part in the simple yet profoundly moving services and to experience something of the joy of the Resurrection in the early hours of Easter morning. Then followed the sharing of bread with a stranger, symbolic of our togetherness which led us to the very heart of the Christian faith.

St Columba landed on Iona from Ireland on May 12th 563 and since those far-off times history has left a number of ancient monuments, Celtic crosses, St Columba's cell, St Oran's chapel. Today one feels thankful to St Columba, who heeded those first spiritual stirrings with such diligence, so that today our souls can be fed in the self-same place as his had been so many centuries ago.

It is always difficult to recapture the essence of such experiences in words, but on my return I tried to do so in the following poem:

Iona of the Dove

O Holy Isle
Iona of the Dove,
To thee a lifelong
Pilgrimage we send
I and a treasured friend.

Another dawn, another day
Have set our feet
Upon this way
And bade our questing hearts
Pursue this isle
Where every hallowed place
Shines with an immortal grace,
Breathes a still sequestered peace.

A gem so rare
Set in a deepening blue,

Agate and turquoise,
Amethyst and rose,
O beauteous land,
Here at thy feet
Forever flows
The white swathe
Of the silver sand,
While from thy breast
The urgent sea recedes
And drawn again rebounds
To take its rest.

This is the place
Long dreamed of,
Wholly fair,
Lifting our eyes
We see Christ
Walking here.
In every human face
His face enshrined,
His peace lays waste
The ravages of time –
Columba's strife.

With forces wild and grim,
The coracle so frail
That carried him
Safely through raging seas
To this calm shore,
To this vast whiteness
Like a threshing floor
Where all is chastened
By the life within.
Spirit of Jesus
Flail our every sin,
O Christ, the life, the grain,
The flail, the floor,
Sift husk from every grain
And in thy mercy pour

This bread upon the hunger
of the world,
Turn water into sacramental wine –
We at thy table dine –
He only lives
Who has the power to love,
O holy isle,
Iona of the Dove.

O bounteous land,
So small, so vast,
So silent after storm
Of ancient days
That ravaged
Thy pure breast
And sought thine
Inmost yearnings
So to desecrate,
Take now thy rest.
For Christ has
All thy ways
Long known and graced
With quietude –
The yearning sands,
The sea-carved rocks,
The fields where cattle
Softly graze,
And ewes who shelter
New-born lambs
Are held within
Christ's weathered hands
And quiet within
This edifice
God's praying people
Find their peace.

O blest Iona,
Island of the Dove,
We offer thee this day

A boundless love
And know that when
From hence we must depart,
We offer thee the Christ,
Within our heart.

37

Italian Pilgrimage

It was at Easter 1979 that another long dreamed-of pilgrimage became possible – this time to Italy, the home of St Francis who for many years had made a deep impression on me. I had read most of the authentic 'lives' and had tried to absorb the spirit of his simplicity as well as his message of oneness with all forms of created life from the highest to the lowest, even as Gandhi and Schweitzer experienced in their search for truth.

Jan and I left London by coach in the early morning and travelled through France via Paris, which looked very much like any other city and rather more shabby than I had remembered it from an earlier visit. The coach stopped every two hours for a brief break and by evening we had reached the Italian Alps, deep in snow, and with snow still falling. When dawn came, we found ourselves travelling between miles and miles of vineyards, tended with great care and considerable industry, for many of the vines had been planted on steep hillsides. There were also many olive groves, the grey and silver leaves of the trees scintillating in the early morning sunlight. Dotted here and there on the terraced hillsides were houses, many without glass in their windows, and on the summit of every sizeable hill there stood a monastery.

Early in the morning we reached Genoa, impressive with its wide boulevards, the famous port from which Americus Vespuccius set sail and discovered what was to become the continent to which he gave his name, America. On descending from the Alps we had driven through endless tunnels penetrating one mountain after another, but the journey was in no way monotonous for sitting on the seat immediately in front of us were two interesting young people on their way to Southern Italy, where a baroness (in deference to the Communists) had offered land to a group of young people.

As the south at that time was very communist, the baroness had decided to share her estate. So for six weeks during the summer these young people ran an inter-cultural seminar on Art, History of Art, Painting, Sculpture etc. They were also constructing a lake. The couple (the husband Serbian, the wife an English artist) wanted to establish a school on similar lines in England. I have never been able to discover if their dream materialised but they, along with the companionship of an English friend who was also unexpectedly travelling in the same coach, added considerably to the pleasure of the long journey.

We finally arrived in Florence, which was our first destination, and from there we made our way by taxi to Fiesole and the Villa de Girolarno, where we were to stay. This is a very ancient monastery with centuries of church history behind it. It is presently run by a nursing order of mainly Irish nuns, but was formerly in the hands of the Jesuits. Now it is a sanctuary of rest and renewal for people needing a period of convalescence after an illness, and as a haven for travellers. The views from the monastery garden overlooking the valley towards Florence were breathtaking, and it was with excited anticipation that we looked forward to discovering some of the most renowned of Florentine treasure during our stay.

We had not however been fully prepared for the intensity of the beauty with which for this brief space our days were to be surrounded. Each day we drank deeply of the atmosphere of this remarkable city and its environs, spread along the banks of the river Arno, as everything we had previously read about its art and architecture suddenly came alive. No wonder that during the middle ages Florence became the intellectual and cultural capital of the whole world. Where else could be found artists of the calibre of Michelangelo, Leonardo da Vinci, Giotto or Titian? And so much creativity inspired by the Roman Catholic faith, for many of the architects were also friars. The city's social conscience was likewise inspired by the same faith, culminating in the provision of refectories for the poor, and care for the mentally handicapped and the orphaned with the establishment of a Foundling Hospital.

Florence was undoubtedly a city of surprises, many of which one came upon unexpectedly. I shall never forget the experience of crossing the Arno via the Ponte Vecchio, or of turning a corner and coming suddenly upon Dante's house or Lorenzo Ghiberti's Gate of Paradise in the Baptistry. Then there were the former homes of the Florentine nobility, the art galleries, countless churches and towering above them all, as if to guard the whole city, its cathedral of St Maria del Fiore. One left Florence with the feeling that a whole lifetime would not be long enough to appreciate all its treasures.

From Florence we travelled to Assisi, in which the life of St Francis is set like a jewel. When we arrived at the convent where we were to stay we found that a good Italian friend, Signora Biondi, to whom we had an introduction via English friends, had welcomed our arrival by putting flowers in our rooms together with cards decorated with pressed flowers from her garden and bearing the greeting *Pax et Bonum*.

Although Jan and I had been overwhelmed by first impressions of Florence, Assisi was equally awe-inspiring in a different way, and one can never forget the first sight of the Basilica, flowing along the ridge of its hill. The first thought was of its incongruity as a memorial to St Francis of Assisi, who gave his clothes away to a beggar and had continually to beg for food. One thing is certain, however, the building of the Basilica is an everlasting tribute to the power of his message.

One is constantly aware that the whole of Assisi speaks of the life of St Francis, as incident after incident leaps out from the pages of history. The crucifix in the half-ruined chapel of St Damian which seemed to speak to Francis, bidding him to rebuild Christ's church, is still preserved by the nuns of St Chiara. The Convent of St Damiano is also an especially sacred shrine, for here St Clare herself lived with her nuns, and the experience of standing before the corner of the dormitory where she died transports one across the centuries to a recollection of her remarkable life.

One is deeply moved too by a visit to the Porgenicula which became the home of St Francis and his followers, and where the saint himself departed this life. But undoubtedly the most inspiring experience of all, and indeed the very heart of our Italian pilgrimage, was to descend into the crypt of the Basilica where the remains of St Francis are at rest in a simple stone casket high above the altar.

After a very impressive Good Friday procession through the streets of Assisi and a service in the Basilica, the following day I was plunged into an experience which I would have preferred to avoid. On walking down the convent's marble stairs I slipped and, falling heavily, fractured my arm and badly gashed my leg. The nuns were shocked and rushed me off to hospital where the leg was duly stitched and the arm encased in plaster. Assigned to a bed in the general ward I prepared to stay there for an unspecified time. This proved to be an interesting experience, for the nursing was done mainly by relatives of the patients. They sometimes appeared in droves, and particularly the men, all dressed up in their Sunday best. In the bed opposite to mine was a terminally ill elderly mother of two kindly middle-aged daughters who in turns kept vigil throughout the day and night. They also cared for me when Jan, who looked after my needs during the day, aided by the nuns at the convent who provided my vegetarian food, had to leave for the night. These new-found friends also gave me an occasional hug to express their sympathy, not knowing any English. Breakfast was hilarious, consisting of a bowl of coffee and a dry bread roll deposited on the bare and not too clean top of each bedside locker. Nothing could have been more Franciscan!

The day's activities were something of a circus, beginning for me with a visit from the doctor who doused the twenty-one stitches in my leg with white spirit, at the same time soaking the bed, while the patients who were recovering and on their feet gathered round my bed to watch. It was all good-humoured and exceedingly friendly.

After a few days, the kindly nuns contacted the hospital

with the request that I should be returned to the convent where they and Jan could continue to care for me and where I would have greater privacy. The request having been granted, I was to undergo the most humbling of experiences, and one of which St Francis would have approved. Four porters loaded me, clad only in my night attire, on to a stretcher, and proceeded to carry me through the streets of Assisi and up the hill – an effort which caused them to lower me on to the ground from time to time while they paused for breath – much to the amusement of numerous passers-by. Then moving slowly on again, they delivered me safely to the convent. There I remained for several days longer until Jan and I were able to make the journey (by plane this time) back to England.

38

India at Last

Ever since meeting Mahatma Gandhi in 1931 I had a deep desire to visit India, but I had to wait almost a lifetime before this became possible, and only then because I could leave my work at the Guildhouse well cared for, and while in India could stay with a cousin, Dora Scarlett, who had set up a medical centre in South India, called Seva Nilayam, meaning House of Service. It was January 1983, and I could have a travelling companion, Ruth Dalton, whose son John had taken over the administration of Dora's leprosy work. Ruth was making one of her fairly frequent visits to India, so it was arranged that I should travel with her.

We met at Heathrow in time for the morning flight and having changed into our summer clothes – incongruously it seemed in the middle of an English winter – the flight path took us over Holland, Yugoslavia, then on to Delhi and Bombay, where we spent the night at the International Airport. The next day we left for Madras where John met us and we had the first real sight of an Indian city. Without a doubt this provided the proverbial culture shock. Hundreds of small shops and pavement shelters jostled, cheek by jowl, with fairly modest and occasionally more palatial houses. The traffic was indescribable.

Ruth and I plus all our baggage were transported on a bicycle rickshaw to the most basic lodge which John had booked for the night. Our accommodation consisted of one small room with three beds and incredibly hard mattresses and pillows. A smaller room adjoining contained very primitive sanitary arrangements, a water tap and a hole in the floor.

It was impossible to sleep and when morning came at last, our breakfast consisted of buns and grape juice with, later on, a cup of tea provided by the Lodge's tea-boy. After breakfast Ruth and I again mounted a bicycle rickshaw and were taken to the home of Peter and Dorothy Millar who were in charge of a large Presbyterian church. Their warm hospitality included a welcome shower and a delicious meal which revived us for the next stage of the journey – eight hours by train to Madurai. Once again we spent a night on the station in a room provided for travellers, amusingly furnished with relics from the days of the British Raj. These included a Victorian hatstand, a very ancient dressing table and eight different electric light fittings, but only two iron bedsteads, so John nobly slept on the floor!

After a good night's rest, we enjoyed a western-style breakfast in the spotlessly clean station restaurant (in contrast to the roads which were often littered with rubbish). We then embarked very thankfully on the last lap of our journey – an eight and a half hours' bus ride to Seva Nilayam. Dora was waiting in the compound to meet us. This was a wonderful moment for us both, as we had not met since we were very young, our lives having taken us geographically in quite different directions. Yet when we talked that evening, we discovered how closely our ideals and aspirations were in tune.

The first impact of the complex of buildings was very impressive: a clinic with a dispensary which caters daily for around two hundred patients, a convalescent unit for mothers and children, and similar residential accommodation for men and boys. Dora's room adjoins a small office which opens on to a veranda from which most of the correspondence and administrative work is carried on. The kitchen and provisions for eating are in a separate building, and the residential helpers have small pleasant rooms in another corner of the compound. There are very basic facilities to meet one's toilet needs, but a good and safe tap water supply, augmented by the installation of a Persian wheel imported by Dora from northern India, along with the labour to install it. To complete the essential buildings are a storehouse for the rice harvest and a simple shelter at the main entrance from the road where an occasional patient can spend the night if necessary. The buildings are set amid a variety of trees including kapok, banana and coconut palms. There are also flowering trees and shrubs, bougainvilleas of many shades, as well as crops of sugar cane, ground nuts, sweet corn and a variety of vegetables. Surrounding the whole are the ripening paddy fields.

The day after we arrived, the air was filled with a kind of subdued excitement as everyone prepared for the celebration of Pongul, a kind of Hindu harvest festival when everyone received a gift. The simple houses were newly painted, as were the bullock carts and the horns of the animals, some of which had one horn coloured orange and the other green, denoting the owner's allegiance to the Congress Party.

On the morning of the festival, numerous beautiful white symbolic designs appeared in many places on the ground. After the Hindu religious services everyone joined in a specially prepared feast – the medical staff, their wives and children, the field workers and the Seva Nilayam patients. The food, which was served on banana leaf plates, consisted of large quantities of rice served with all kinds of cooked vegetables, sauces and sweet rice with ghee – all vegetarian. It was a happy and colourful scene with the women in their brightly coloured saris, often with flowers in their hair, and the children likewise arrayed in their new clothes. The following day it was the animals' turn to have a celebration. They were first bathed with soapy water and if their horns were not painted, they were oiled. Then each animal was decorated with flowers in readiness for a special ceremony in the afternoon, after which they were given sweet rice, bananas and sugar cane, of which everyone present also received a small share.

After Pongul, life settled down to a more ordered routine. The clinic was staffed by a doctor, a number of paramedical helpers – most of whom lived outside – and from time to time a nurse came from England. The staff arrived very early, as did their patients, often carrying a sick child. It was a long, exacting day, but to lift even a little of the weight of human suffering brought its own reward.

Another aspect of Dora's work which impressed me

deeply was the small leprosy hospital called Arogam Agam. Situated a little distance away from Seva Nilayam, this was sensitively administered by John Dalton with whose mother, Ruth, I had travelled to India. There I saw patients with the disease in various stages of severity being treated with the utmost care and concern. Mercifully leprosy can now be arrested, but often irreparable physical damage has been done before medical aid is sought.

To facilitate the treatment of leprosy patients further afield, a jeep equipped as a clinic and accompanied by John Dalton and several paramedical workers makes regular visits to the outlying villages. On one occasion I had the privilege of accompanying them to Alagapari where, under the shade of a tamarind tree, leprosy patients of all ages gathered for treatment. This was a moving experience when one realised that in this act of compassion lay their one hope of healing.

Other memories crowd into my mind as I write. On one occasion Dora, always eager to show me more of India, suggested a journey to Kodai Kanal. This involved a long, uphill journey through the most spectacular countryside, a veritable tapestry with brightly coloured flowering trees, the road then rising slowly above mountain peaks to 7000 feet. Kodai was formerly a hill station to which missionaries retired during the hot season, so in some respects it remains partially westernised and still accommodates a number of Roman Catholic schools. We stayed overnight in an almost empty bungalow, incongruously named Daisy Bank, which only boasted some very ancient beds, a rickety table and one or two distinctly precarious chairs.

The following day we travelled on to Munnar, and from there we climbed to 8229 feet on the highest road in India south of the Himalayas. Then down again to Kumily and Periyar House with its fascinating nature reserve for wild animals and exotic birds. The return journey followed the mountain path through Kerala, with its mile after mile of tea plantations covering the valleys and high on the hillsides in meticulous formation.

Kerala itself is the second richest state after the Punjab,

and evidence of its prosperity is reflected in its gaily painted houses and public buildings, as well as in its lush and well cared-for countryside. It claims to have been converted to Christianity by St Thomas the Apostle, and subsequently continued by missionaries of the Roman Catholic faith. This religious persuasion, still in existence today, enables it to have a more liberal attitude to inter-caste marriages than that which prevails in many other states. Generally, throughout India, ninety per cent of marriages are still arranged. Couples do not marry primarily for love but to have children, especially an all-important son who can eventually perform his father's funeral rites by cremation. Occasionally, when fuel is unobtainable, burials take place in the fields with the touching injunction, 'Come to the fields and go home'. Happily, love often develops even in arranged marriages when both partners accept marriage together as their destiny.

The return journey through Kerala to Tamil Nadu in which Seva Nilayam is situated left us with a host of memorable impressions: like a rich tapestry embodying so much beauty of the countryside with its plantations of teak, eucalyptus and castor oil trees, and an undergrowth of herbs and spice-producing shrubs. These were interspersed with forests of flowering trees, magnificently arrayed in varying shades of green, pink, red, light mauve and cream – a veritable Garden of Eden. And there was life everywhere. In the forests and in the fields and villages, the brightly coloured saris of the women enhanced the landscape as did the beauty of the smiling children with whom one could always make friends so easily, and for whom a difference of language was never a barrier. When we could not speak, we sang together and sometimes even danced for joy!

A few days later, as my stay in India was coming to an end, Dora thoughtfully arranged another expedition, this time to visit her friend Father Bede Griffiths at the Satchidananda Ashram at Shantivanam. I had expressed a wish to meet him, for in his early days Father Bede had lived in Gloucestershire and I had followed his spiritual pilgrimage with great interest. This time we were able to make the journey by bus, arriving at the ashram in the late afternoon. Its numerous attractive buildings are set among palm trees and gardens, surrounding a central compound with seats for open-air gatherings. There is a temple, a circular meditation hall, a library, a large airy room for meals, sleeping accommodation for staff and guests as well as a number of individual cells for theological students who attend courses there. The life is ordered and peaceful, with prayers or mass at regular intervals throughout the day. When we were there Father Bede gave talks in the evening on the Bhagavad Gita, with its insights into right living.

Following a directive from the Vatican, members of Catholic communities in India are advised to refrain from attempting to convert to Christianity adherents of the other great faiths, cultivating instead the common ground of their beliefs and incorporating into the mass Hindu symbols such as camphor and sandalwood. In spite of all the missionary endeavour over the years, only two per cent of the total population of nearly 900 million Indians claim to be members of the Christian church. Consequently, the Vatican Council, in the Declaration on Non-Christian Religions, declared that 'the Church rejects nothing that is true and holy in these religions' and encouraged Catholics to 'recognise, preserve and promote the spiritual and moral values as well as the social and cultural values to be found among them...'

Among the gifts given by God to India, the greatest was seen to be that of interiority, the awareness of the presence of God in the heart of every person and of every creature, and which is fostered by prayer and meditation, by contemplative silence and the practice of Yoga and Sannyasa. 'These values' it was said, 'belong to Christ and are a positive help to an authentic Christian life.' Consequently, it was said, 'Ashrams where authentic Christian spirituality is lived, should be established, which should be open to non-Christians, so that they may experience genuine Christian fellowship.'

Before leaving Shantivanam we were taken to see a Sarvodaya spinning centre established in a former

Brahmin's house, which also accommodated a small soap-making unit, typical of many such cottage industries to be seen throughout the length and breadth of India. After a final talk with Father Bede we made our way back to Seva Nilayam, mentally stimulated and spiritually enriched by the experience.

Now, a number of years later, the whole of the Indian experience remains vividly in mind and I shall never cease to be thankful to Dora and everyone else involved who made it possible for me to feel, even for a brief space, the deep heartbeat of Gandhi's beloved India.

At the conclusion of my story I wish to thank all those who have contributed to the building and the work of the Stanton Guildhouse by their labour, their gifts, their friendship and their prayers. Without them the Guildhouse would not exist.

To them, and to those who will continue its work in the future, I dedicate this book.